Environmental Change
and Globalization

Environmental Change and Globalization

Double Exposures

ROBIN M. LEICHENKO AND KAREN L. O'BRIEN

UNIVERSITY PRESS
2008

OXFORD
UNIVERSITY PRESS

Oxford University Press, Inc., publishes works that further
Oxford University's objective of excellence
in research, scholarship, and education.

Oxford   New York
Auckland   Cape Town   Dar es Salaam   Hong Kong   Karachi
Kuala Lumpur   Madrid   Melbourne   Mexico City   Nairobi
New Delhi   Shanghai   Taipei   Toronto

With offices in
Argentina   Austria   Brazil   Chile   Czech Republic   France   Greece
Guatemala   Hungary   Italy   Japan   Poland   Portugal   Singapore
South Korea   Switzerland   Thailand   Turkey   Ukraine   Vietnam

Published by Oxford University Press, Inc.
198 Madison Avenue, New York, New York 10016

www.oup.com

Oxford is a registered trademark of Oxford University Press

Library of Congress Cataloging-in-Publication Data
Leichenko, Robin M.
Environmental change and globalization : Double exposures/
Robin M. Leichenko and Karen L. O'Brien.
   p. cm.
Includes bibliographical references and index.
ISBN 978-0-19-517731-2; 978-0-19-517732-9 (pbk)
1. Global environmental change. 2. Globalization.
I. O'Brien, Karen L. II. Title.
GE149.L45 2008
304.2'5—dc22   2007039764

9 8 7 6 5 4 3 2 1

Printed in the United States of America
on acid-free paper

For
Annika, Charles, Espen, Henry, and Jens Erik

Acknowledgments

We have been working on this book project for a number of years, and we are grateful to many people and institutions for helping us along the way. Our former PhD advisers, Diana Liverman and Amy Glasmeier, first encouraged our collaboration over a decade ago and have continually supported our work. The U.S.–Norway Fulbright Foundation sponsored Robin Leichenko's research visit to Norway, where the book project began. Our colleagues and students in the Department of Geography at Rutgers University, the Department of Sociology and Human Geography at University of Oslo, and CICERO provided support throughout the project. We also benefited from the advice and encouragement of William Solecki, Kristian Stokke, Elizabeth Wentz, Karen Arabas, Janine Stenehjem, Rodney Erickson, and Lynn Rosentrater.

Many colleagues and students have played an important role in our thinking about the interactions between global environmental change and globalization, and in the development of the double exposure framework. Our collaborators on the India research, including Guro Aandahl, Stephan Barg, Suruchi Bhadwal, Akram Javed, Ulka Kelkar, Bhujang Rao, Heather Tompkins, Henry Venema, and Jennifer West, helped us to initially "ground truth" the double exposure concept. William Solecki, Siri Eriksen, Julie Silva, Hallie Eakin, Neil Adger, Coleen Vogel, Mike Brklacich, Hans Georg Bohle, Gina Ziervogel, Jon Barnett, Mark Pelling, Tom Downing, Richard Klein, Bronwyn Hayward, Colin Polsky, and many others have inspired us through their research, writing, and commentary. The students at the International Human Dimensions Programme (IHDP) capacity-building workshop on Global Environmental Change, Globalization and Food Systems in 2004, and the IHDP workshop on Vulnerability in the Context of Globalization in 2005 gave us valuable comments on early versions of the material.

We are especially grateful to Ken Conca and Hallie Eakin, who provided detailed and thoughtful comments on the full manuscript, and to Kirsten Ulsrud and Øystein Kristiansen for useful editorial suggestions. Marthe Stiansen gave us enormous help with the figures and maps. Finally, we thank our parents, Eleanor and Harold Leichenko and Barbara O'Brien, and our families, Henry and Charles Strehlo and Jens Erik, Espen, Annika and Kristian Stokke, for their continuous love, good humor, and support.

Contents

Environmental Change
and Globalization

Introduction

1

The nature of changes now occurring simultaneously in the global environment, their magnitudes and rates, are unprecedented in human history and probably in the history of the planet.
—W. Steffen et al., _Global Change and the Earth System: A Planet Under Pressure_

Few areas of social life escape the reach of processes of globalization. These processes are reflected in all social domains from the cultural through the economic, the political, the legal, the military, and the environmental.
—David Held et al., _Global Transformations: Politics, Economics and Culture_

Introduction

We live in an era of profound change. The melting of permafrost and ice in the Arctic, the disappearance of mangroves and wetlands, loss of biodiversity, and reductions in levels of stratospheric ozone represent just a few of the environmental changes that pose unprecedented risks and threats to individuals and communities. At the same time, changing patterns of production, the spread of mass consumerism, commodification of water and other natural resources, and proliferation of new communication technologies are facets of a larger process of globalization that presents both substantial challenges and significant opportunities to different regions, sectors, and social groups. While each of these changes may appear to be

3

isolated and unrelated, we demonstrate in this book that they are closely inter-twined, creating situations of "double exposure" for many regions, communities, and individuals.

Although humans have been transforming the Earth for millennia, it is only recently that these cumulative processes and impacts have been framed as global issues. From a historical perspective, both global environmental change and glo-balization may be viewed as reflections of the long-term expansion of human settle-ment and use of the natural environment (Turner et al. 1990; Diamond 1999, 2005; Merchant 2003; Clark et al. 2005). Large-scale environmental changes began with the earliest civilizations as land and water resource systems were transformed to facilitate the spread of organized agriculture (Worster 1985; Cronon 1991; Diamond 1999; Elvin 2004). While commerce and trade among different groups also date back thousands of years, the emergence of an international market economy can be traced to the network of private European trading corporations established more than 500 years ago (Polanyi 1944; Crosby 1972; Kindleberger 1996; Hirst and Thompson 1999).

What is *different* about global change processes in the twenty-first century is the speed, scale, and extent to which they are altering biophysical, social, and ecological systems (Beck 2000; Speth 2003; Steffen et al. 2004; Friedman 2005; Homer-Dixon 2006). What is *disconcerting* about these processes is their dramatic yet unequal consequences for humans and other species. Although both global-ization and global environmental change are sometimes characterized as univer-sal and all pervasive, the literature on each process is replete with references to "winners and losers," "included and excluded," or "privileged and marginalized" (Conroy and Glasmeier 1993; Glantz 1995; Kapstein 2000; Khor 2001; O'Brien and Leichenko 2003; O'Loughlin et al. 2004; Dicken 2007). Uneven outcomes from global environmental change manifest via changing temperature patterns, changes in water availability, reduction of species habitat, and loss or gain of livelihood opportunities. Inequalities from globalization are reflected in uneven rates of eco-nomic growth and capital mobility, polarization of income and wages, differential access to political power, and limited diffusion of new technologies.

Global environmental change and globalization reflect issues of equity (Roberts and Parks 2006; O'Brien and Leichenko 2006; Held and Kaya 2007; UNDP 2007), but they are also about uncertainty. The rate, timing, and magnitude of environ-mental change will depend on future patterns of energy use, population growth, and urban settlement, which are difficult to project with any certainty. The pace and pattern of globalization are similarly unpredictable, particularly in light of technological change and political dynamics. The uncertainties surrounding each process often manifest through unexpected shocks and stresses. Droughts, floods, sea level rise, and toxic algae blooms may occur in connection with global environ-mental change, while industrial plant closures, currency devaluations, and price spikes for oil and other commodities may result from globalization. These types of extreme events place a premium on the capacity to respond and adapt to change (Barnett 2001a). Such events often contribute to new types of vulnerabilities among individuals, households, communities, and regions, and they may also undermine the resilience of social and ecological systems (Downing 1991; Dow 1992; Adger and Kelly 1999; Pelling 2003a; Skoufias 2003; Turner et al. 2003a; O'Brien et al. 2004; Aggarwal 2006; Kirby 2006; Polsky et al. 2007).

Still another hallmark of both global processes is the notion of growing connectivity (Homer-Dixon 2006; Sklair 2002; Beck 2000). Many understandings of global environmental change emphasize the linkages and feedbacks between the biophysical and human components of the Earth system, and most interpretations of globalization highlight the economic, political, cultural, and technological integration that is increasingly linking distant places. Through these connections and linkages, actions taken in one part of the world, or by one group, have increasingly visible effects on other locations and groups. Present actions may also have increasingly significant impacts on future generations. These spatial and temporal linkages often create feedbacks that accelerate rates of change. A key concern is that these interacting processes are surpassing the capacity of humans and ecosystems to adapt, thereby undermining long-term sustainability (Beck 1992, 2000; Sklair 2002; Steffen et al. 2004; Schellnhuber et al. 2004; Lovelock 2006; Dicken 2007).

Global environmental change and globalization are transformative processes with significant consequences for social and environmental systems. Yet the linkages and connections between the two processes are often obscured by separate discourses on each topic. The two processes are rarely studied together, and research and policy discussions on each process tend to be highly compartmentalized, focusing on narrowly bounded topics such as trade liberalization, climate change, land use change, or international migration. Within these separate discourses, the critical linkages, feedbacks, and synergies between globalization and global environmental change often go unnoticed.

In this book we present a framework for analyzing the interactions between global environmental change and globalization. The double exposure framework emphasizes that there are multiple types of interactions between the two global processes, which we refer to as "pathways" of double exposure. These pathways often lead to growing inequalities, increasing vulnerabilities, and accelerating rates of change. Yet the interacting processes also present openings for responding positively to change. By drawing attention to double exposure, we show that global environmental change and globalization can together create new opportunities and synergies that may enhance human security.

━━━━━━ *Global Environmental Change*

Global environmental change has been largely framed as an issue that is distinct from globalization. It is typically defined as a set of changes to the Earth system that are expected to have major effects on human society and ecosystem services. Most of these changes can be attributed to human activities, which have increased in magnitude, extent, and tempo (Steffen et al. 2004). Global environmental change includes both systemic changes that directly influence how the Earth system functions, and cumulative local changes that influence the global environment through the magnitude and distribution of their effects (Turner et al. 1991). Thus the "global" aspects of environmental change refer to both the scale and extent of the changes. Examples of systemic changes that are planetary or global in scale include increasing concentrations of greenhouse gases, stratospheric ozone depletion, and sea level rise. Cumulative changes—also described as local events that occur worldwide and affect a significant share of total global resources—include water pollution, land use changes, and soil degradation (Turner et al. 1991; Pelling 2003b).[1]

Most prominent among the systemic changes are those associated with the atmosphere, particularly climate change or global warming. Climate change scenarios produced by general circulation models of atmosphere-ocean-ice dynamics project dramatic consequences to result from increasing atmospheric concentrations of greenhouse gases such as carbon dioxide, methane, and nitrous oxide (Karl and Trenberth 2003). Observed and projected changes in climatic conditions are expected to lead to unprecedented changes in ecological and social systems. These effects may include increased frequency of drought and water stress in many regions, shifts in species habitat and greater risk of species extinctions, disruption of human settlement patterns, particularly in coastal areas, increased heat-related mortality, and significant alterations in global food production patterns (Dow and Downing 2006; IPCC 2007a; IPCC 2007c).

The consequences of climate change will be widespread, but they will not be equally distributed (Parry et al. 2001). Greater temperature increases are expected in higher latitudes and inland areas than in tropical regions and coastal areas. Although precipitation changes are more difficult to model and project and therefore more uncertain, most scenarios indicate a general trend toward drier conditions in already dry areas and wetter conditions in moist areas (IPCC 2007c). Global sea levels are projected to rise by as much as 0.5 to 1.4 meters (compared with 1990 levels) by 2100 (Rahmstorf 2007). Model results are, of course, dependent upon scenarios of greenhouse gas emissions, and most often they represent changes in average conditions. Nevertheless, it is widely recognized that such changes will be driven or accompanied by changes in climate variability and the magnitude or frequency of extreme weather events such as hurricanes and tropical cyclones (Mearns et al. 1997; IPCC 2007b). Climate variability, influenced in part by El Niño-Southern Oscillation (ENSO) and its teleconnections, is also considered to fall within the realm of global environmental change, particularly in relation to droughts and floods.

Prominent cumulative environmental changes include land use change and changes in marine ecosystems (Schneider 2004; Foley et al. 2005; Berkes et al. 2006). Estimates of the amount of human-induced changes to land cover worldwide suggest that approximately half of the ice-free land surface of the Earth has been substantially affected by human activities (Steffen et al. 2004). These changes, which include wetlands loss and deforestation, have significant implications for ecosystem services (Millennium Ecosystem Assessment 2005). Tropical deforestation in particular has raised concerns about a growing loss of biodiversity, as well as the implications for local and regional climates (O'Brien 1998; Curran et al. 2004). Human activities have also led to changes in the abundance, diversity, and productivity of marine populations (Berkes et al. 2006). The loss of marine bio-diversity is increasingly impairing ecosystem functions and reducing the ocean's resilience to perturbations (Worm et al. 2006).

Habitat changes account for many of the threats to biodiversity.[2] Yet the intentional and unintentional introduction of exotic or invasive species through trade and global transportation systems also endangers biodiversity, particularly endemic species on small islands. Furthermore, exploitation of species through extractive industries and illegal trade threatens many species of flora and fauna. On top of these threats, climate change will have also widespread consequences for bio-diversity and may endanger species that are otherwise not facing risk of extinction (Araújo and Rahbek 2006). Overall, human actions are creating new conditions that have implications for all types of species and ecosystems. Of the 1.5 million identified species, one in three amphibians, including almost half of the world's freshwater turtles, are considered endangered, as well as one in eight birds and one in four mammals. In fact, over 16,000 species are listed as endangered in the IUCN's Red List of Threatened Species (IUCN 2006).

Concerning other cumulative changes, humans play a particularly important role in the global water cycle and have a strong influence on the local availability and quality of water resources. Human intervention in the global water cycle can be tied to urbanization, industrialization, and hydropower development, which have led to the damming, diverting, and draining of rivers with a wide array of hydrological impacts (Worster 1985; Conca 2006). Although many of these transformations are only visible at the local scale, their cumulative effects may lead to changes in the global water system that surpass the impacts of recent and anticipated climate change over decadal time scales (Vörösmarty 2002).

Human actions are considered to be a driving force behind global environmental change, whether through population growth, resource extraction, energy consumption, urbanization, technological change, changes in consumer demands, or shifts in attitudes, lifestyles, and values (NRC 1992; Steffen et al. 2004). Central to global environmental change studies is the idea that human activity is changing natural, life-supporting processes. Environmental changes in turn affect human societies through impact on human lives, infrastructure, resources, and ecosystem services (Millennium Ecosystem Assessment 2003, 2005).

Globalization

Globalization is often understood as a movement toward greater economic, political, and cultural integration across nations (Held et al. 1999; Sklair 2002; Dicken 2007). Globalization is commonly associated with "space-time" compression, whereby improvements in transport and communication technologies have reduced the time and costs associated with the movement of goods, people, and ideas across space, effectively "shrinking" the world (Harvey 1990). Most definitions also emphasize the growing connections between people and groups all over the world (Beck 2000; Sklair 2002; Homer-Dixon 2006). These changes are reflected in many different globalization-related transformations, including rising levels of international trade and investment, formation of transnational commodity chains, increasing travel and transnational migration, expansion of communication networks, the emergence of a global mass media, liberalization of financial markets, and homogenization of consumer cultures across a growing "global" middle-income class.

While all of these changes serve to enhance integration among individuals, firms, and nations, there is widespread recognition that patterns of globalization are highly uneven and that many regions and groups are marginalized by these processes (Mittelman 1994; Held and McGrew 2002; Dicken 2007; Held and Kaya 2007). Differential levels of involvement in international trade, varying access to media, communication and transportation networks, and differential patterns of foreign investment and technological diffusion all reflect the uneven nature of globalization (Sklair 2002; Leichenko and Silva 2004; Knowles 2006). The outcomes of globalization for individuals, communities, social groups, and regions are also highly uneven. Growing polarization of wages and income levels both within and across nations, increasing inequalities in access to health and social services, and differential access to channels of political power are frequently connected to processes of globalization (Rodrik 1997; O'Loughlin et al. 2004; Silva and Leichenko 2004; Held and Kaya 2007).

Globalization is frequently associated with the ideology of neoliberalism, which advocates free-market approaches to economic activity (i.e., letting markets operate unfettered by state regulation). This ideology emphasizes growth, productivity, and market efficiency, affecting policy choices and decisions at all levels, from international forums to promote free trade, such as the World Trade Organization (WTO), to national and subnational efforts to attract foreign direct investment and privatize state-owned enterprises, to regional and local measures promoting decentralization of political power, local capacity building, and other forms of "New Public Management" (Peck and Tickell 2002; Leichenko and Solecki 2005). Neoliberal ideology also influences political involvement at the local level by stressing the role of citizens as autonomous, individual consumers rather than as members of a political community (Klinenberg 2002; Slocum 2004).

Despite general agreement about the major facets of globalization, there remain prominent debates about the character, scope, and significance of these changes. Much of the contemporary academic literature on globalization emphasizes questions of what globalization is, whether or not it is "real," and how it is different from prior periods of capitalist expansion in the nineteenth century and earlier (Amin 1997; Hirst and Thompson 1999; Bridge 2002; Sklair 2002; Speth 2003; Dicken 2007). Some view globalization as part of a long-term process of "creative destruction," whereby "old ideas and organizations are constantly challenged and replaced by new, better ones" (Rajan and Zingales, 2003, 1). Whereas earlier phases were associated with countries and companies globalizing, the most recent phase can be characterized as globalization at the level of individuals and small groups, including diverse and non-Western individuals who are taking advantage of revolutionary changes in information and communication technology (Friedman 2005). Within the current phase, globalized social movements not only have become a powerful mechanism for responding to change, but also are changing the very nature of globalization (Kingsnorth 2003; McDonald 2006).

The implications of globalization for the environment have not gone unnoticed (Haas 2003; Speth 2003; Barnett and Pauling 2005; Zimmerer 2006). Globalization of beef consumption is seen as a driver of both deforestation and biodiversity loss in the Amazon (Nepstad et al. 2006), and a growing level of urban affluence is recognized as a significant contributor to greenhouse gas emissions and climate change (McGranahan 2007). Globalization has also been linked to a loss of biodiversity: "Rather than encouraging local diversity, the globalization of culture, foodways, languages, export crops, and exotic invasive species ensures that these local elements are replaced by exotic imports with wide distributions" (Redford and Brosius 2006, 317). Even the growth in public awareness of global environmental issues and new forms of environmental governance at all levels has been linked to globalization (Held et al. 1999; Zimmerer 2006). Nevertheless, within both academic and policy realms, globalization is generally viewed as separate and distinct from global environmental change, with each process having its own set of driving factors, each creating uneven outcomes, and each requiring different types of policy responses. Framed as discrete processes, they reveal two separate pictures of a changing world in the twenty-first century. Yet global environmental change and globalization may also be seen as linked processes that interact to create double exposure.

Double Exposure

In film photography, double exposure describes the accidental creation of images, where the intention was not to photograph over a previously exposed image but to produce two separate pictures. The result is an unintended and often blurred image that reflects the merged outcome of both shots. Double exposure is also an artistic technique whereby film is exposed to one subject and then, before being processed, is exposed to another subject. The artistic double-exposed photograph is deliberately created to achieve a desired outcome. In either case, the viewer perceives the double-exposed shot as a single picture and responds to that new image rather than to its constituent parts. In this book, double exposure is a metaphor for cases in which a particular region, sector, social group, or ecological area is simultaneously confronted by exposure to both global environmental change and globalization (O'Brien and Leichenko 2000). As with photography, double exposure to global environmental change and globalization leads to outcomes that may be accidental or intentional and may be viewed as either negative or positive. Whatever the case, the interactions between these processes are creating new contexts that call for not only new responses but new ways of looking at the two processes.

This exploration of double exposure begins with an examination of the separate discourses on global environmental change and globalization. Chapter 2 argues that the big picture is obscured by separate and relatively narrow discourses and framings of the two processes. Chapters 3 and 4 move beyond the metaphor of double exposure to present a more detailed framework for investigating the interactions between global environmental change and globalization. The framework draws on research in both areas and presents a synthetic approach to investigating processes, outcomes, and responses to global change. When considered together, the two interacting processes challenge the ways that global environmental change and globalization are currently viewed and addressed, and at the same time they reveal possibilities for creating new responses to change. Although many of the detailed empirical examples presented here draw upon our research in the area of climate change, the double exposure framework provides a generalized approach for examining interactions between many facets of global environmental change and globalization.

Double exposure often entails being vulnerable to negative outcomes that result from some combination of stresses and shocks related to *both* processes. In other words, those experiencing the negative effects of environmental change often simultaneously feel the negative effects of globalization. For example, in 1997 the co-occurrence of both a labor market shock tied to the devaluation of the Thai baht and an El Niño-related drought shock affected 19 percent of the population surveyed by Datt and Hoogeveen (2003) in the Philippines. The adverse impact of double exposure increased with the level of education, ownership of land, and level of the community's commercial development. At the same time, those individuals, groups, or regions that are positively affected by one process often tend to be positively affected by the other process. Estonia, for example, is likely to experience increased forest productivity under climate change, which may have positive consequences for timber exports (Nilson et al. 1999). These overlapping negative and positive effects help to explain how and why the exposure to the two processes may contribute to growing social and economic inequalities and polarization. These types of overlaps are illustrated in more detail in chapter 5, which shows how some regions, communities, and households involved in agriculture in India are double exposed to both climate change and trade liberalization.

Double exposure is more than simply the overlapping outcomes of globalization and global environmental change. The term also describes the ways that the two processes influence exposure and capacity to respond to a wide array of stresses and shocks. Shocks may include naturally occurring events, such as earthquakes, as well as those tied more directly to environmental and economic changes, such as floods or currency devaluations. Stresses include changing rainfall patterns, the large-scale shift of jobs to other parts of the world, and increasing coastal erosion. Global environmental change and globalization are, in other words, creating new contexts for experiencing and responding to change. Belliveau et al. (2006), for example, describe how Canadian farmers responded to increased competition from foreign wines by planting premium European varieties of grapes. These new varieties proved to be sensitive to climate stresses that decrease wine quality, thus creating a new type of vulnerability for wine producers. As chapter 6 shows, new vulnerabilities are also appearing in urban areas as the result of both processes of global change. A case study of Hurricane Katrina in New Orleans illustrates how both processes contributed to a context for disaster.

Adopting a longer-term and more dynamic perspective on double exposure makes it clear that the two processes are closely interrelated. Contextual changes and outcomes resulting from one process may lead to responses that drive the other process, creating feedbacks that accelerate change and influence future outcomes. Berkes et al. (2006) describe the new dynamics of marine resource exploitation that result from globalization. In particular, new global markets may develop so quickly that local institutions are unable to respond to the rapid pace of exploitation. For example, an increased global demand for the green sea urchin in the 1980s led to waves of exploitation that masked regional depletion, accelerating declines in this fishery. Such feedback reveals both contradictions and synergies between responses to each process that can lead to accelerating change. As is discussed in chapter 7, responses both to climate change-related reduction of sea ice and to the development of international shipping in the Arctic are likely to facilitate resource extraction, which may be a driving force behind climate change in the future.

Double exposure emphasizes the interactions between two transformative processes of change. Through the double exposure framework, we highlight those interactions between the two processes that are contributing to growing inequalities, increasing vulnerabilities, and accelerating and unsustainable rates of change. The double exposure framework not only sheds light on dangerous consequences that may be masked by separate framings of the two processes, but also reveals possibilities for using the interactions to generate outcomes associated with more positive visions of society.

These positive visions are encapsulated in notions of human development, well-being, and human security (Barnett 2001b; Commission on Human Security 2003; Lister 2004; Gasper 2005; McGrew and Poku 2007; UNDP 2007). Human security, in particular, describes a condition whereby individuals and communities have the options necessary to end, mitigate, or adapt to threats to their human, social, and environmental rights, and where they have the capacity and freedom to exercise these options (GECHS 1999). Human security encompasses more than traditional notions of state security, which emphasize safety from armed conflict and violence, and it expands upon the notion of environmental security, which looks at how resource scarcity, environmental degradation, and environmental stress can influence both conflict and cooperation (Barnett 2001b). Rather, human security relates to the well-being of individuals and incorporates issues related to human capabilities, human rights, and environmental sustainability (Commission on Human Security 2003; Gasper 2005; Dodds and Pippard 2005; Liotta and Shearer 2006). By emphasizing the opportunities presented by the interactions between the two processes, we show that it is possible to identify new openings and entry points for increasing equity, strengthening resilience, and enhancing sustainability, all of which contribute to greater human security.

Conclusion

Global environmental change and globalization are two overarching processes that are weaving together the fates of households, communities, and people across all regions of the globe (Held and McGrew, 2002). Nevertheless, surprisingly little attention has been given to the full range of interactions between the two processes. Researchers and scholars of each process tend to operate within their own separate discourses, focusing on some interactions while ignoring others. Furthermore, there is a tendency within both realms to focus on narrowly bounded processes rather than on their dynamic relationship with outcomes and responses. This approach reflects how science is socially organized, with emphases on new discoveries and theories over applied research. In many cases, "processes themselves become the object of explanation and are themselves regarded as outcomes" (Bridge, 2002, 362). The emphasis is placed on understanding global environmental change and globalization as two separate phenomena rather than on investigating the interconnections between processes, outcomes, and responses.

One of the aims of this book is to expand the window of analysis for research on global change to incorporate the multiple linkages and interactions between global environmental change and globalization. By understanding and addressing the dynamics of these two interlinked processes, we can begin to develop a better picture of the challenges and opportunities for enhancing human security in the twenty-first century.

Global Change Discourses

2

...we have to live with the fact that different individuals and groups use different discourses to make sense of the same nature/s. These discourses do not reveal or hide the truths of nature but, rather, *create their own truths*. Whose discourse is accepted as being truthful is a question of social struggle and power politics.
—Noel Castree, "Socializing Nature: Theory, Practice, and Politics" (emphasis in original)

Global environmental change and globalization each emerged as a major scientific research area in the mid- to late 1980s. Their concurrent appearance is not coincidental: both processes were pushed into the forefront of scientific interest by major world events, including the break-up of the Soviet Union and the end of the Cold War in 1989, publication of the Brundtland Commission report on Our Common Future (WCED 1987), and the first major public warnings about the dangers of climate change (Hanson 1988). Each of these events served to reinforce public and scientific recognition that fundamental, global-scale, structural changes were happening. The fields of global environmental change and globalization have since matured and thereby have also fragmented into a myriad of research communities.

The separation of these communities within academic research on global change is striking. Much of the research on global environmental change is carried out within the physical and biological sciences, with the goal of understanding the complex processes underlying various aspects of global environmental change, including the atmosphere, oceans, and cryosphere, as well as ecosystem dynamics (Steffen et al. 2004). The societal impacts are often regarded as an outcome to be studied once the "science" is better understood and the physical impacts have been identified (Wynne 1994). Although there is a strong emphasis on interdisciplinary environmental research, perspectives on global environmental change from the social sciences and the humanities have been difficult to integrate with many biophysical studies (Newell et al. 2005). A resulting scientific division of labor between the biophysical sciences and the social sciences not only influences how the issues are understood but also places the two in competition with one another for access to research funding and political influence (Pelling 2001, 172).

In contrast to research on global environmental change, studies of globalization primarily fall into the realms of social science and humanities, including disciplines such as geography, economics, sociology, anthropology, political science, history, philosophy, and comparative literature (e.g., Sassen 1998; Held et al. 1999; Sklair 2002; Stiglitz 2002, 2006; Tabachnick and Koivukoski 2004; Dicken 2007). Globalization studies are also well represented in the fields of business management and law. Although some globalization research addresses environmental issues, much of this work is limited to questions about how globalization affects environmental quality and environmental management. This literature does not typically consider how it influences or interacts with larger processes of environmental change.

In this chapter we consider different understandings and interpretations of each process of global change. Drawing from discourse analysis and theories about the social construction of scientific knowledge (e.g., Samantha Jones 2002; Marsden et al. 2002; Adger et al. 2001; Livingstone 1992), we stress that understandings of global processes are embedded in specific and often competing sets of discourses. A discourse can be defined as a specialist language that describes the world in a particular way, making certain claims to truth and justifying certain types of knowledge and certain forms of action (Foucault 1977; Escobar 1996; Peet and Watts 1996). Discourses include assumptions, values, judgments, and contentions that provide the basic terms for analyses and debates, influencing both agreements and disagreements (Dryzek 2005). A discourse may also be understood as a "system of representation made up of rules of conduct, established texts and institutions which regulate what meanings can and cannot be produced" (Smith 1998, 254).

Through an examination of the discourses on each process, we show that "global environmental change" and "globalization" can be interpreted in different ways. Importantly, discourses carry political weight and reflect underlying power structures, which give currency and legitimacy to some viewpoints over others and which maintain the interests of some over the well-being and security of others. In addition to power relations, discourses also reflect worldviews, including the basic assumptions and beliefs that influence much of an individual or group's perceptions of the world, their behavior, and their decision-making criteria (Kearney, 1984).[1] Whether conditioned by religion, gender, political affiliation, or social class, worldviews influence how people make sense of and respond to change (O'Riordon and Jordan 1999). As will be shown below, discourses not only interpret what the concepts mean, but also control how they are used, how they influence the questions that are asked, and how they prioritize the level or scale of analysis, all of which ultimately have implications for strategies to address global change (Forsyth 2003).

────────   ## Discourses on Global Environmental Change

We argue in this book that global environmental change represents a set of trans-formations that interact with globalization through a number of pathways. As such, we adhere to a broad "global" discourse on environmental change, which holds that contemporary environmental problems are not bounded by location, and that the global nature of environmental issues today demands new approaches and strategies. We also see ecological risks as increasingly internationalized, with threats and hazards emerging independent of the place where they are produced (Beck 1992; Millennium Ecosystem Assessment 2005).

It is important to recognize that this global discourse is contested and that there are skeptics who question the existence and severity of global environmental problems, particularly climate change (e.g., Lomborg 2001; Taylor 2005; Lindzen 2006). Skeptics use statistics and scientific argumentation to show that environmental change is not taking place, or to argue that it is not overwhelmingly negative. They often criticize the global view as being propounded by environmental evangelicals, apocalysts, doomsdayers, or naysayers, arguing that development has led to vast improvements in the environment for many and that environmental quality will further increase with development and economic growth. The skeptical view argues against action to mitigate the drivers of global environmental change, charging instead that global environmental change is a fiction that is used to justify interventions in the economy (or in specific industries) as part of a "green agenda,"[2] or that regulatory actions are unlikely to be cost-effective and that technological solutions promoted through a free market are more promising (Switzer 1997). We will return to this skeptical view on global environmental change in a later discussion of discourses on globalization.

This section focuses on discourses that accept global environmental change as a real and pressing issue. It shows that, embedded within the broad global discourse, there are a number of ways of understanding global environmental change, each of which is based on a particular approach to science. It identifies three key subsets of the global discourse: the biophysical discourse, the human-environment discourse, and the critical discourse. These subsets are not intended to be mutually exclusive, nor are they necessarily internally homogenous and consistent. Within a particular discourse there may be heated theoretical debate about some issue or relationship, but each subset nonetheless captures some of the critical assumptions behind both research and policy agendas for global environmental change. The three subsets are also not intended to be comprehensive and all-inclusive; other discourses and clas-sifications of discourses can be identified (see Adger et al. 2001; Dryzek 2005). The objective here is to show that there are very different ways of understanding and addressing global environmental change.

The Biophysical Discourse

The first and most prevalent discourse on global environmental change is the "bio-physical discourse," which frames environmental problems as biophysically-based concerns that need to be addressed through international commitments to reduce, limit, or suspend activities that lead to environmental change. Research within this discourse views changes in biogeochemical cycles, atmosphere-biosphere-ocean interactions, and other natural and physical processes as the most important determinants of outcomes. Mathematical models are often used to develop projections of temperature change, sea level rise, acidification of precipitation, the extent of the ozone hole, land use change, and so forth. The biophysical discourse emphasizes that we have entered a new period in Earth history—often referred to as the "Anthropocene era"—in which human activities play a decisive role in influencing natural systems (Steffen et al. 2004).

Adopting a holistic worldview, proponents of the biophysical discourse often take the dynamic Earth system as a starting point and investigate how human activities are affecting biological and physical conditions and processes (Schellnhuber et al. 2004). The Gaia hypothesis, which sees the Earth as a self-regulating system, akin to a living organism, has also contributed to interdisciplinary earth systems science and the biophysical discourse on environmental change (Lovelock 1988; 2006; Kineman 1991).[3] The discourse places particular emphasis on the ways that human activities, such as fossil fuel consumption and forest clearance, modify the system and its critical components. Demographic, social, and economic trends and "carrying capacity" are thus viewed as critical to determining future impacts on the environment, particularly as they relate to greenhouse gas emissions, land use change, water supply and quality, and biodiversity (Vörösmarty et al. 2000; Tilman et al. 2001; Myers and Kent 2003). Nonetheless, social science input on issues of global environmental change is usually limited to the information on population changes, economic growth and development, and technological change that future scenarios require (Pitman 2005).[4] The structural and sociospatial changes associated with globalization are seldom included within this discourse.

Addressing environmental problems from the biophysical perspective involves, first and foremost, understanding Earth system processes and how these processes are influenced by human activities. The biophysical discourse is based on the belief that if scientific knowledge can be developed enough to reduce uncertainty, then a basic social consensus on actions and responses will follow. The discourse calls for better science, including "significant advances in basic knowledge, in the social capacity and technological capabilities to utilize it, and in the political will to turn this knowledge and know-how into action" (NRC 1999, 160). Such ideas are often propounded by international organizations, international scientific networks, environmental NGOs, and the media. Proponents of this discourse are especially visible within international scientific programs charged with studying environmental change, such as United Nations Environment Programme (UNEP) and the International Geosphere Biosphere Programme (IGBP). For example, the widely cited Millennium Ecosystem Assessment (MA) emphasizes the consequences of changes in ecosystem services for human well-being, both at present and into the future (Millennium Ecosystem Assessment 2003).[5] The goals of the MA are also consistent with the notion that decision makers need a better scientific understanding of environmental conditions and that this improved knowledge will translate into policy actions to address environmental problems (Millennium Ecosystem Assessment 2003).

The biophysical discourse on global environmental change is also prominent in the realm of international environmental politics, in which international environmental problems are approached through numerous treaties and regulatory regimes (Switzer 2004). International environmental agreements include efforts to discontinue production of ozone-depleting chemicals, to slow deforestation, to limit emissions of greenhouse gases, to regulate trade in endangered species, and to manage wetlands.[6] Influencing environmental agendas and political responses, the biophysical discourse can thus be considered a powerful discourse on global environmental change. Nevertheless, there is a growing recognition that humans not only drive environmental change but are also unevenly affected by its consequences, a factor that further influences the environment. Over the past decade, attention to both human and biophysical dimensions of environmental change has resulted in an increasingly visible "human-environment" discourse on global environmental change.

The Human-Environment Discourse

The human-environment discourse emphasizes the linkages between social and physical systems. Proponents of this discourse argue that humans are an integral part of the environment, or part of a "coupled social-ecological system" that is characterized by continual interactions between physical and social processes. Rather than seeing the biophysical and social as separate and distinct systems, this discourse views the natural environment as being inseparable from human activities (Berkhout et al. 2003; Easterling and Polsky 2004; Clark et al. 2005). The human-environment discourse responds both to critiques of research that focuses exclusively on biophysical aspects of global environmental change, and to critiques of research that focuses only on social perspectives.[7]

There are two major, overlapping subcategories within the human-environment discourse: those that focus on contextualized understandings of human-environment systems (e.g., Kasperson et al. 2001; Turner et al. 2003a) and those that draw on ecological perspectives and systems theory, with an emphasis on coupled social-ecological systems (e.g., Holling 2004; Walker and Salt 2006; Young et al. 2006). Although these two subcategories emerged from different intellectual backgrounds,[8] both invoke a similar language of vulnerability, resilience, and adaptation (Adger et al. 2003; Folke et al. 2002; Polsky 2004; Schröter et al. 2004; Polsky and Cash 2005, Berkhout et. al. 2006; Arvai et al. 2006; Berkes 2007). Both approaches also recognize that the global environmental system has entered a human-dominated epoch and that there is a need for holistic rather than discrete analyses of ecological and social systems.

Research approaches within the human-environment discourse emphasize the importance of integrating both biophysical and social vulnerability in order to understand how outcomes of environmental change are distributed within society. Proponents of this discourse also place strong emphasis on adaptation and social learning as responses to environmental change (Thompson et al. 2006; Clark et al. 2005; Holling 2004). Examples of successful adaptations may include technological solutions and innovations such as developing seeds that tolerate drier conditions, building dikes or setbacks that limit damage from storm surges, or establishing captive breeding programs for endangered species. Institutional adaptations may entail changing strategies for water management by involving markets to encourage efficient responses. Adaptive management, collaborative learning, and information sharing at the international level are promoted as promising responses to global environmental change (Crabbe and Robin 2006; Mendelsohn 2006; Thompson 2006). Related avenues for responding to environmental change include the formation of partnerships among governments, private-sector actors, civil society, and academia. Some research within the human environment discourse also recognizes that facets of globalization, such as liberalization of international trade, may influence vulnerability and resilience to global environmental change (e.g., Adger 2000a, 2000b; Leichenko and O'Brien 2002; Eakin 2006; Young et al. 2006). These and other types of interactions between the two global processes are discussed in more detail in later chapters.

The human-environment discourse is widespread among the human dimensions community on global environmental change research. It has been described by Kates et al. (2001) and others as a "new science of sustainability," and it is beginning to gain currency within large international organizations such as the United Nations (UN) and World Bank. Similar to the biophysical discourse, there is an assumption that "society lacks a critical understanding regarding which kinds of programs, institutional arrangements, and more generally, 'knowledge systems' can most effectively harness [science and technology] for sustainability" (Cash et al. 2003, 8086). Integrative approaches to human-environment systems tend to adopt an implicit view of politics that emphasizes democratic process, transparency, and open public debate. Like the biophysical discourse, the human-environment discourse subscribes to the idea that more and better scientific understanding of the drivers, processes, outcomes and responses associated with global environmental change is what is needed and that, once this information is provided, the appropriate actions will be taken to reduce negative effects.[9] These underlying assumptions of both the biophysical and human-environment discourses have been questioned and challenged in what we describe as a critical discourse on global environmental change.[10]

The Critical Discourse

Whereas the biophysical and human-environment discourses emphasize the *environmental* aspects of global environmental change, critical voices argue that the focus should more appropriately be placed on the political, economic, moral, and cultural dimensions (Wynne 1994). Within the critical discourse, particular attention is given to the role of human beings as social agents, capable not only of influencing change but also of responding in a variety of ways to shape outcomes, either positively or negatively. On the basis of structural and political analyses, adherents of the critical discourse emphasize that the outcomes of environmental change are not determined by physical and ecological factors alone and that wider social, economic, and political issues must be addressed before environmental problems will be solved. Critical approaches thus recognize that transformative changes associated with globalization are influencing environmental and social outcomes (e.g., Eaton 2003; Bebbington 2001).

Numerous approaches to understanding global environment change are included within the critical discourse. Here we will draw attention to two approaches, political ecology and ecofeminism, in order to illustrate the diversity of views within the critical discourse. Political ecology approaches explore how economic, political, and social relations influence the framing and understanding of environmental problems and issues (Blaikie and Brookfield 1987; Robbins 2004). Peluso and Watts (2001, 18), for example, challenge the way in which environmental change is framed in studies of environmental security, in particular the "implicit recognition that all environmental processes generate shortage and scarcity." They point out that many of these studies fail to acknowledge how forms of capital accumulation intersect or ally with specific forms of power to produce environmental degradation.

Ecofeminist perspectives emphasize that environmental problems stem from the interrelated domination of women and nature (Eaton and Lorentzen 2003). These approaches stress the need for elimination of male-gender bias in the production of knowledge and advocate interpretations that recognize that life in nature, which includes human beings, is maintained through cooperation, mutual care, and love (Eaton and Lorentzen 2003; Mellor 2003; Mies and Shiva 1993). Mies and Shiva (1993), for example, criticize the notion of promoting equality through "catching-up development," which involves emanating capitalist-patriarchal models that degrade people and the environment. They point to contradictions in pursuing sustainable development through commodity production and technological fixes, and they propose a subsistence perspective whereby economic relationships are aimed toward satisfying fundamental human needs, based on new relationships to nature and among people, carried out through participatory democracy.

Proponents of the critical discourse also argue that a focus on the global scale and on biophysical and ecological processes draws attention away from the social, political, and gender relations that shape these processes, including the scientific interpretations of the processes. The critical discourse is thus closely linked to social theory and poststructuralist and postmodernist approaches to knowledge, including critiques of the Enlightenment paradigm of positivist science (see Smith 1984; Redclift and Benton 1994; Castree 1995, 2001). Drawing on the work of philosophers and postmodern theorists, researchers working within this discourse argue that scientific inquiry, theories, and hypotheses represent social constructions that are influenced by history and by the current cultural, political, and economic context (Samantha Jones 2002; Marsden et al. 2002). Castree (2001, 3) emphasizes the importance of power and interests in these social constructions: "nature is defined, delimited, and even physically reconstituted by different societies, often in order to serve specific, and usually dominant, social interests."

The critical discourse also reflects "postnormal science," which emphasizes the need for much wider participation in the science-policy process to address issues of differing perspectives, interests, and uncertainty (Ravetz 2005). Not surprisingly, the critical discourse has been less visible within global change research and policy circles, including within international negotiations on issues related to global environmental change. This lower profile can be partially attributed to the perception that concepts and ideas within this discourse are presented in a language that is intelligible only to those familiar with the work, which makes it difficult for a general audience to grasp meaning and significance (Schneider 2001). Another perception is that the work downplays the significance of environmental change processes, favoring structural arguments over environmental facts and diminishing the importance of biophysical changes (Samantha Jones 2002).[11] Notwithstanding these perceptions, the critical discourse draws attention to the fact that important issues are at stake in debates over the construction of nature, both political and philosophical (Demeritt 2001). Attention to politics and differential power within the critical discourse captures some of these interests and conflicts that are often ignored in the biophysical and human-environment discourse (see Taylor and Buttel 1992).

---   *Discourses on Globalization*

As with global environmental change, there are many discourses surrounding globalization. Although there is general agreement about basic components of globalization, those working within each discourse disagree about more fundamental issues, including whether globalization is generally beneficial or detrimental, how to respond to differential outcomes of globalization, and what types of research issues and policy responses are most salient. There is also a separate dialogue about the "reality" of globalization—with one side taking a hyperglobalist position that globalization is a recent and all-pervasive process, and the other taking the "skeptical position" that globalization is not a new phenomenon but simply a continuation of long-term trends toward modernization and expansion of global capitalism (see Wallerstein 2000; Held and McGrew 2002; Dicken 2007). Prominent skeptics include Hirst and Thompson (1999), who use empirical evidence about global economic trends to show that the current global economic regime is less open and less integrated than the one that prevailed from 1870 to 1914. Skeptics also suggest that "globalization" represents a discourse that is used by those in power to further certain agendas, such as reducing national labor protections and giving tax breaks to multinational corporations, and to paralyze efforts aimed at national and local economic management—all of which are justified in the name of being globally competitive (Hoogvelt 1997; Hirst and Thompon 1999).

Rather than delving further into the debate about the reality of globalization, we will focus here on three major discourses that accept that globalization is a significant and fundamentally transformative process. We again recognize that these groupings are not fixed or definitive (see Rupert 2000; Held and McGrew 2002) but instead provide insights into why interpretations and actions surrounding globalization differ. The first two discourses reflect the opposing "public faces" of globalization (Bridge 2002; Milanovic 2003), which are often evident in polarized debates, namely, benign globalization and malignant globalization.[12] Between these polarized positions are discourses that seek to alter or reshape the present form of globalization. These discourses are combined here into the category of "transformative globalization," which includes both enthusiasts and critics of globalization, all of whom share a desire to make globalization more equitable in terms of the distribution of power, opportunities, and life chances (Held and McGrew 2002). These categories of benign, malignant, and transformative globalization are intended to capture key differences between those who favor the current form or trajectory of globalization, those who strongly oppose most aspects of globalization, and those who seek a very different form of globalization. As with the categorizations of global environmental change discourses, there are often significant disagreements within each discourse, but at the same time there is general agreement about how globalization is defined and whether and how it might be addressed.[13]

Benign Globalization

The benign globalization discourse presents globalization as a positive and benevolent force that will lead to greater prosperity through economic growth, through increasing production efficiency, and through the transfer of technology (Milanovic 2003). This discourse, which has been referred to as the "Washington Consensus," "the neoliberal position," and most recently as the "Post-Washington Consensus," advocates continued liberalization of economic policies in order to promote open markets, free trade in goods and services, and free movement of capital (Held and McGrew 2002; Onis and Senses 2005). Particular emphasis is placed on the gains from open trade, which include economies of scale in production, emergence of new types of jobs, and growing access to new technologies, ideas, and consumer goods (Ohmae 1995; Friedman 2000; Dollar and Kraay 2002). As is noted by Milanovic (2003, 667), this view also implies that globalization will lead to "converging institutions as democracy becomes the global norm, and cultural richness as people of different backgrounds interact more frequently."

Policies associated with the benign discourse emphasize avenues through which to foster processes of globalization, including opening national borders to international trade via lower tariffs, deregulating industry, and privatizing government services. The creation of "friendly" or "attractive" climates for foreign direct investment is emphasized; these often involve institutional reforms, guarantees of tax relief, and measures to ensure good governance (Rajan and Zingales 2003). Another prescription promoted by international financial institutions such as the International Monetary Fund, the World Bank, and regional development banks has been a series of tax reforms, with the understanding that a reduction in receipts brought about by reduced marginal tax rates and diminished trade taxes would be countered or exceeded by an expansion of the tax base as well as by better administration and enforcement of tax collection (Sanchez 2003).

The benign globalization discourse not only stresses the economic and political benefits of globalization but also emphasizes the benefits for environmental quality. Various rubrics and frameworks, including the environmental Kuznets curve and ecological modernization, are employed to make the case for a positive relationship between globalization, improvements in technology, and environmental quality (see McCarthy 2004; McGranahan 2007). Concerning environmental governance, private property–oriented solutions are considered optimal over regulatory or state-centered approaches. Examples of free-market approaches to environmental governance include programs for tradable emissions permits, user fees for public goods such as parks, privatization of water supplies and other public utilities, and reductions in state subsidies for agricultural production. All of these efforts are consistent with the tenets of neoliberalism, which emphasizes free markets, deregulation, a rollback of state intervention into private-sector activities, and private-sector management of public goods (Liverman 2004; McCarthy 2004; McCarthy and Prudham 2004; Zimmerer 2006). Notably, the arguments about the benefits of globalization for the environment are largely consistent with the position of the skeptics of global environmental change, who suggest that economic growth offers a way to solve environmental problems (Jacques 2006).[14]

Proponents of the benign globalization discourse include not only those who advocate neoliberal, free-market viewpoints but also those who desire some degree of managed control over these free-market processes via treaties, policies, and international institutions in order to promote more equitable distributional outcomes (Onis and Senses 2005; Stiglitz 2006). In recent years there has been growing recognition among economists that markets are imperfect and that neoliberal globalization is likely to produce a number of losers (Kay 2004). Although these are generally thought to be losers only in the short term, it is increasingly acknowledged within the benign discourse that globalization requires management (Onis and Senses 2005). Consequently, many operating within the benign discourse advocate a weak form of global management of the economy, as well as short-term "adjustments" to help those who are harmed by globalization.

The benign discourse is evident in many of the contemporary public discussions of globalization, particularly within the media, major international institutions such as the World Bank and International Monetary Fund, and the international business community. This discourse is also dominant in much of mainstream economics and political science literatures (Milanovic 2003). Because the benign discourse is so prevalent, it is often invisible to those who accept and use it. One factor that explains its predominance among "experts" is the use of a "language of inevitability" (Gilbert 2005).[15] Where the boundaries and assumptions of the benign discourse become most evident is when those operating outside the discourse present alternative viewpoints and question the logic of free markets and economic openness.

Malignant Globalization

The malignant globalization discourse views globalization as the unrestrained expansion of capitalism and the American business model, which is leading to an internationally sanctioned pillaging of resources and the destruction of local livelihoods and environments (Shiva 2000; Khor 2001; Speth 2003). Within this discourse, globalization is seen as primarily a mechanism of exploitation, which benefits certain vested interests and, in the process, destroys local cultures, weakens labor protection, undermines local environments, and threatens the political and economic autonomy of both countries and regions (Anderson 2000; Shiva 2002; Speth 2003, International Forum on Globalization 2002; Ritzer 2004). Economic changes that promote integration are interpreted as an expansion of corporate control that benefits members of the transnational capitalist class but that disadvantages the poor and working classes (Amin 1997; Greider 1997; Shiva 2002). Given its highly critical stance, the malignant discourse is often characterized as a "radical" position (Held and McGrew 2002).

Within the malignant discourse, globalization is associated with, among other things, the disappearance of social safety nets, the loss or relocation of jobs in industrialized countries to create "sweatshops" in developing countries, the forced transnational migration of women, and the destruction of rural livelihoods (Khor 2001; Ehrenreich and Hochschild 2002; Shiva 2002). Culturally, global corporations— and global capitalism more generally—are seen as having largely unchecked power to shape societies (Greider 1997). Globalization is also considered as a homogenizing force that erases cultural differences and leads to the replacement or absorption of local cultures by a global culture (Ritzer 2004). Proponents of the malignant discourse often view globalization as synonymous with Americanization and the spread of the least desirable elements of American popular culture, including "the hotdog, the hamburger and the handgun" (Kirby 2004, 134).[16]

The negative environmental consequences of globalization are also a concern within the malignant discourse. The discourse stresses that globalization increases the influence of market forces on the formulation and enforcement of environmental policy, and at the same time subjects national environmental policies to the demands of international economic institutions (Zarsky 1997). Market-driven economic globalization also pushes countries toward convergence in environmental policy. Indeed, the possibility that globalization can exert downward pressures on national environmental standards has been widely discussed and debated. Issues related to transparency and democracy in decision-making, indigenous knowledge and Trade-Related Aspects of Intellectual Property Rights (TRIPS), biotechnology, and consumer and worker rights as trade barriers have also formed the basis for strong critiques of globalization and its consequences for the environment (Nader et al. 1993, Speth 2003).

Other voices within this discourse advocate responses aimed at limiting or reversing the present course of globalization in order to protect local economies and environments. Advocates of "localization" challenge globalization by arguing for alternative modes of development (Curtis 2003; Hines 2003). As summarized by Hines (2003, 4):

At the heart of localization is the replacement of today's environmentally and socially damaging global subservience to international competitiveness.

In its place it prioritizes local production and the protection and rediversification of local economies and environments, such that everything that can sensibly be produced within a nation or region should be.

Various measures to achieve localization include use of locally based currencies, selective use of trade protectionism and policies to promote local production, and ecological taxes on both producers and consumers (Hines 2003; Curtis 2003; Princen et al. 2002).

Proponents of the discourse of malignant globalization are often visible in protests and demonstrations against efforts to perpetuate liberalization of trade policies and other forms of international economic integration. Because this discourse reflects opposition to many of the policies advocated by proponents of the benign (and dominant) globalization discourse, its proponents, including members of various social and labor movements and NGOs, often have little voice within major international organizations and the business community. Labeled by the mainstream media as "anti-globalization" or "anti-WTO" (i.e., World Trade Organization), these movements interpret globalization as a fundamental threat to nationally based social and environmental protections. Globalization is perceived as a force intended to usurp the power of the nation state by making the state subservient to the demands associated with open markets (Peck and Tickell 2002). Beck (2000, 1) provides an apt summary of this position:

> The premises of the welfare state and pension system, of income support, local government and infrastructural policies, the power of organized labour, industry-wide free collective bargaining, state expenditure, the fiscal system and 'fair taxation'—all this melts under the withering sun of globalization and becomes susceptible to [demands for] political moulding.

Transformative Globalization

The discourse of transformative globalization adopts a more nuanced definition of what globalization is, viewing it as a multiscalar set of processes operating at global, national, regional, and local levels (Mittelman 2000). This discourse, which is notably present in the geographic and other social science literatures, interprets globalization as a set of processes that not only influence events and activities at the regional and local levels, but also are driven by local- and regional-level actions and decisions (Bebbington 2001; Held 2004; Lee and Yeoh 2004; Dicken 2007). As Lee and Yeoh (2004, 296) describe it, "[G]lobalisation itself is materially and discursively constructed out of complex interactions and power struggles between diverse social, political and economic actors occurring simultaneously, rather than hierarchically, across various geographical scales." Structural transformations associated with globalization are also seen as erasing traditional North–South divisions and creating a new "social architecture" that cuts across national and geographic boundaries (Castells 1998; Hoogvelt 1997; Held and McGrew 2002).

Those operating within the transformative discourse acknowledge that the benign (i.e., neoliberal) form of globalization is currently the dominant form, and they seek more socially and environmentally just alternatives. Beck (2000), for example, interprets globalization as entailing a reduction in the power of the national state and an increase in interdependencies between places. He sees globalization as creating new relations of power and competition, and new conflicts and intersections between national states and new transnational actors, identities, social spaces, and processes (Beck 2000). As with the malignant discourse, Beck's version of the transformative discourse takes a position that equates globalization with the expansion of capitalism and corporate control and the loss of power of nation states to regulate their economies and to provide social insurance.

Sklair (2002, 8) considers the dominance of transnational practices controlled by a transnational capitalist class as a key facet of contemporary globalization. This class sees its economic interests as global rather than national or local in origin; it includes the owners and controllers of major multinational corporations, as well as government bureaucrats, politicians, and professionals.[17] Transnational practices, which again have economic, political, and cultural elements, entail new ways to organize social life across existing state borders, emphasizing a shift in the balance of power between state and nonstate actors and agencies. Transnational corporations and the transnational capitalist class are seen as the key locus for transnational economic and political activities, while the proliferation of a culture-ideology of consumption (which is spread through both transnational corporations and the transnational capitalist class) is seen as a key dimension of transnational culture. Sklair (2002) proposes an alternative form of globalization that entails a shift toward socialized globalization, led by self-governing, transnational communities and replacement of the culture-ideology of consumption with a culture-ideology of human rights, which emphasizes meeting basic human needs of all people.

Many proponents of the transformative discourse stress the opportunities that arise from globalization, especially through political and cultural changes such as democratization, reform of the UN system, greater awareness of the value of cultural differences, and formation of new international social networks (Held 2004). The discourse also recognizes that globalization has the potential to offer many benefits via better communication and new forms of community that are enabled via technologies. At the same time, it recognizes that the present form of globalization needs to be fundamentally altered in order to promote both process and outcomes that are more equitable and sustainable (Sklair 2002; Onis and Senses 2005).

An important commonality among proponents of the transformative discourse is an emphasis on the linkages between globalization, environmental degradation, and social polarization (Sklair 2002; Dicken 2007). Proponents of the transformative discourse emphasize that globalization should be more sustainable and socially equitable, although there are differing views on the appropriate avenues for change. They also emphasize that domestic policies remain essential for growth and development, whether through promotion of domestic savings, implementation of countercyclical fiscal policies, mobilization of public resources, investments in education, promotion of employment, or reduction of income inequality (Sanchez 2003). Another prescription for transformation of the current form of globalization is a project of global social democracy, as described by Held (2004). Held's approach advocates new and more equitable modes of transnational cooperation based on the rule of law, political equality, democratic process, and social justice.[18] Others within the transformative discourse see globalization as a call for new normative and conceptual understandings of democracy, which Bohman (2007) refers to as "transnational democracy."

Alternative forms of engagement with globalization are also emphasized within the transformative discourse. Mittelman (2000), for example, points to the spread of resistance as a global phenomenon that shapes globalization itself. Bebbington (2001) similarly illustrates how local communities may take advantage of multiple sites of resistance via connection with various transnational networks. These sites allow local actors to participate in alternative types of globalized social and economic relationships with the dual goal of promoting livelihood security and improved environmental quality. For example, the promotion of bottom-up globalization that emphasizes new alliances between nature and culture and new forms of local democracy can be seen in many "self-sustainable local development" initiatives (Magnaghi 2000).

Given the challenges that the transformative discourse poses to the dominant form of globalization, it is not surprising that it has a limited voice within mainstream media, policy, and business communities. Nonetheless, proponents of this discourse convey a vital message that globalization outcomes are in no way inevitable, thus making room for both national policies and local action to influence the processes (Rodrik 1997; Cox 1997; O'Brien and Leichenko 2003).

Conclusion

Whether framed from a contemporary or longer-term perspective, the phrases "global environmental change" and "globalization" foster images of large-scale events that influence everyone and everything. Images of planet Earth are used to symbolize global interconnections; slogans such as "think globally, act locally" implore people to consider their activities in a global context; and expressions such as "our common future" and "the global village" imply that humankind has a common destiny. By focusing on different discourses on global change, we have drawn attention to the fact that "global environmental change" and "globalization" mean different things to different people or groups. Although there are common themes and buzzwords associated with each process, a myriad of voices are looking at the processes and their outcomes, and a "global" view can be interpreted in many different ways. Furthermore, the various discourses are not equal: some are given more weight and power within formal institutions while others are silenced or subdued. Researchers and actors operating within specific discourses are often unaware of how their discourse may be used to limit policy options or support the pursuit of particular interests or agendas.

Although the need for multiple viewpoints and interdisciplinary approaches to research is increasingly emphasized and prioritized, particularly in research on global environmental change, academic and policy communities are not always willing to acknowledge other discourses, nor are they always open to the inclusion of different perspectives. This tendency to operate within a single discourse not only limits the types of research questions that are investigated, but also misses critical interactions *between* the two processes. The next two chapters present a double exposure framework that draws attention to interactions, linkages, and feedbacks between the two processes that are often overlooked in the separate discourses on global environmental change and globalization. By identifying multiple pathways of interaction, the framework shows how processes together can exacerbate inequalities, increase vulnerabilities, and undermine sustainability, thus posing significant challenges to human security. Recognition of these interactions can, in turn, reveal new openings and opportunities for responding positively to change.

Double Exposure: A Conceptual Framework

3

It is difficult to be in the midst of change and to understand it:
to separate the deep structures from the shallow constructions,
the long rhythms from time's embellishments, or the profound
understandings from the current enthusiasms. And we are very
much in the midst of change.
—Kates et al., "The Great Transformation"

The global change discourses illustrate a variety of approaches toward understanding the dramatic transformations taking place in the world today. In this chapter we shift our focus from discourses on global environmental change and globalization to frameworks for research and analysis. Research frameworks provide conceptual and methodological tools for exploring specific types of physical and social phenomena. Because research frameworks are typically embedded in specific discourses, they often provide limited views on global change processes, missing many types of interactions and linkages between global environmental change and globalization.

The double exposure framework presented here provides a conceptual tool for investigating the interactions between global environmental change and globalization. Such interactions are at the heart of many contemporary issues, ranging from the use of biofuels as a mitigation response to climate change and its implications for food security to the protection of marine species in the face of expanding global markets. The framework shows how the two processes are continually transforming the context in which individuals, regions, communities, and social groups experience and respond to change. Emphasizing the dynamic and progressive nature of the two processes, the framework raises critical questions about human security that are often ignored in the separate discourses on each process:

- Who is negatively or positively affected by both global environmental change and globalization? Why do the uneven outcomes of the two processes frequently overlap? What are the implications of these differential outcomes for issues of equity?
- How does globalization influence vulnerability to global environmental change? How does global environmental change affect the capacity to respond to globalization? What are the options for increasing resilience in the face of multiple processes of global change?
- In what ways does globalization contribute to global environmental change, and vice versa? To what extent do these linkages and feedbacks lead to accelerating rates of change? What are the implications of these interactions for efforts to promote sustainability?

By demonstrating how the processes overlap and interact, the framework captures the multidimensional character of the changes taking place under the separate labels of "global environmental change" and "globalization."

Global Change Frameworks

A myriad of research frameworks have been used to investigate processes and outcomes of both global environmental change and globalization. We categorize these frameworks into those that emphasize processes of global change, impacts of global change, or vulnerabilities to global change. These categories are not meant to be exhaustive. Rather, they illustrate some of the major ways that each process has been approached, thereby providing a backdrop for the presentation of the double exposure framework.

Process Frameworks

Process-oriented frameworks emphasize description and explanation of dynamics of both types of global change. These frameworks are widely used in global environmental change research, particularly within studies rooted in the biophysical discourse. They are designed to explain phenomena such as climatic change, ozone depletion, and changes in global water cycle, taking into account both anthropogenic drivers and the biophysical feedbacks within the Earth system (Steffen et al. 2004). The Integrated Model to Assess the Global Environment (IMAGE), for example, represents a comprehensive framework for examining changes in the atmosphere, oceans, cryosphere, and other facets of the Earth system, focusing on their causes and links (Bouwman et al. 2006).

Process-focused frameworks are also widely used with studies of globalization and its various dimensions such as trade liberalization, integration of communication networks, and changing institutional structures. The OECD Handbook on

Globalization Indicators, for example, presents a framework for analyzing the economic, technological, commercial, and financial dimensions of globalization and for constructing internationally comparable indicators that measure the magnitude and intensity of globalization (OECD 2005). The Foreign Policy Globalization Index takes into account economic integration, technological connectivity, personal contact, and political engagement to measure which countries are and are not globalizing (Kearney 2005). While the OECD and Foreign Policy approaches are rooted largely in the discourse of benign globalization, process-oriented globalization frameworks appear within all of the major globalization discourses (Bridge 2002).

Impacts Frameworks

Impact-oriented frameworks emphasize how exposure to processes of global change affects specific regions, sectors, and social groups. These frameworks, which are generally associated with the biophysical or human-environment discourses on global environmental change or the benign or malevolent discourses on globalization, detail how processes such as climate change, ozone depletion, trade liberalization, or financial deregulation affect various outcome metrics. These metrics may include, for example, crop production, water availability, sectoral employment, human health, and poverty rates (e.g., Rosenzweig and Solecki 2001, Leichenko and Silva 2004; O'Loughlin et al. 2004; Huynen et al. 2005). A prominent example of global environmental change impacts framework is the IPCC Technical Guidelines for Assessing Climate Change Impacts and Adaptations (Carter et al. 1994). This widely used framework employs climate change scenarios generated by general circulation models to assess how changes in temperature, precipitation, and other physical parameters are likely to affect different regions or economic sectors. The IPCC framework has also been expanded and modified to incorporate uncertainties associated with climate change scenarios as well as spatial and temporal variations in human adaptation (Jones 2001; Polsky 2004).

Many types of impacts frameworks are employed to assess social, economic, and other effects of globalization. The work of Huynen et al. (2005) illustrates an impacts framework for studying how human health is affected by different facets of globalization, such as increasing cross-border mobility, changes in lifestyles, and new methods of food production. Other impact frameworks are used to assess the effects of globalization on economic conditions such as employment, wages, and income (e.g., Bernard and Jenson 2000; Silva and Leichenko 2004). These frameworks each develop indicators of exposure to import competition, exchange rate volatility, or other measures of globalization and then explore how changes in exposure affect wage levels, sectoral employment, or income distribution. Regardless of the outcome metric, all impacts frameworks share an emphasis on evaluation of the consequences of a specific facet of global change for individuals, households, or other exposure units.

Vulnerability Frameworks

Vulnerability frameworks explain how and why some groups and individuals experience negative outcomes from global change shocks and stressors (Schröter et al. 2004; Kirby 2006). Such frameworks often focus on the specific contextual factors that influence exposure and the capacity to respond to change (Turner et al. 2003a; Wisner et al. 2004; Ionescu et al. 2005; Luers 2005; Eakin and Luers 2006; Eriksen and Kelly 2007; O'Brien et al. 2007). For example, Turner et al. (2003a) developed a place-based framework that focuses on the coupled human-environment system and examines how hazards can potentially affect the system. Their framework recognizes that responses and their outcomes collectively determine the resilience of the coupled system and may, in fact, transcend the system or location of analysis to affect other scalar dimensions of the problem, creating potential feedbacks to the original system.

Related frameworks focus primarily on the notion of resilience, emphasizing how social and ecological systems respond to perturbations of all types (Gunderson and Holling 2002; Walker and Salt 2006). Resilience frameworks identify those characteristics that allow human, natural, and coupled social-ecological systems to effectively respond, adapt, or recover from various shocks with minimal disruption, and promote adaptive co-management as a strategy for dealing with complex systems (Ruitenbeek and Cartier 2001; Plummer and Armitage 2007). Characteristics that contribute to resilient social-ecological systems include the capacity to live with uncertainty and change, maintenance of diversity in its various forms, the ability to combine different types of knowledge for learning, and the opportunity for self-organization and cross-scale linkages (Berkes 2007).

Other types of vulnerability frameworks include capabilities, assets, and livelihoods approaches that focus on the factors that constrain or enable people in pursuing outcomes that they value (Moser 1998; Sen 1999; DFID 1999; Wood 2003). The DFID (1999) framework on sustainable livelihoods, for example, views people as operating in a context of vulnerability, where they have access to certain assets or poverty-reducing factors that are influenced by the prevailing social, institutional, and organizational environment. While vulnerability frameworks are rooted largely in the human environment discourse, they often draw on insights from the critical and malevolent globalization discourses, including perspectives from political economy, political ecology, environmental justice, and other bodies of literature that emphasize the role of class, gender, and structural inequalities in contributing to a context of vulnerability (Kirby 2006).

In addition to an emphasis on vulnerability or the capacity to respond to single processes of global environmental change or globalization, several frameworks explore differential vulnerabilities to *multiple stressors* (O'Brien et al. 2004; Luers et al. 2003; Turner et al. 2003a).[1] Multiple stressor approaches recognize that individuals and communities are often subject to more than one process of global change. Within the human-environment discourse in particular, recognition of multiple stressors is seen as key to understanding why some regions, individuals, or groups are more capable or less capable of adapting to global environmental change (O'Brien and Leichenko 2000; O'Brien et al. 2004). These approaches typically combine investigation of vulnerability to changing environmental conditions with examination of other stressors such as disease, warfare or conflict, and economic disruptions (Eakin 2006; Eakin at al. 2006; Belliveau et al. 2006; Lind and Eriksen 2006; Ziervogel et al. 2006). Expanding the analyses beyond "single stressor" approaches has provided greater insights on social processes that shape vulnerability and responses to global change processes.

What's Missing?

Although all of the types of frameworks just described provide important insights regarding processes, outcomes, and responses to global environmental change and globalization, several important limitations should be noted. First, the frameworks do not take into account the full extent of the interactions between the two global change processes. Global environmental change and globalization both intersect and interact in multiple ways to influence processes, outcomes, and responses over time. For example, the coincidence of currency shocks and El Niño-related weather shocks, as noted in chapter 1, not only may differentially influence household incomes but may also change the context for responding to future events through the erosion of social or community capital, or it may create feedbacks to both globalization and global environmental change processes by triggering migration to urban areas (see Datt and Hoogeveen 2003; Hecht et al. 2006). Although it is increasingly recognized that people and places are experiencing manifestations of both processes at the same time, methodologies and frameworks for analysis of these interacting processes are still lacking.

Second, there has been relatively little recognition within existing frameworks of how the two processes *together* transform the context in which people and places experience and respond to changes of many types. Many of the global change frameworks stress the importance of context for explaining both differential outcomes and vulnerabilities (see Lambin et al. 2001; Turner et al. 2003a; Pelling 2003a; Perrons 2004; Wisner et al. 2004; Eakin 2005). For example, the Pressure and Release (PAR) model of Wisner et al. (2004) explicitly discusses how "unsafe conditions" are transformed into disasters given exposure to both biophysical and socio-political-economic stress. Yet these frameworks seldom recognize the extent to which the context itself is dynamic, dramatically changing as the result of both global environmental change and globalization (Leichenko and O'Brien 2002; Eakin 2006). A crop failure in India, for example, can slow the economy and spur supply-side inflation while at the same time putting pressure on the government to address food insecurity. In this dynamic context, monetary policies are increasingly affected by market-determined interest rates, which are closely tied to climate conditions (Mukherjee 2005). These changing contextual conditions affect exposure and responses to future global change processes, thus creating new patterns of vulnerability and new challenges for social and ecological resilience (O'Brien et al. 2007).

Finally, most frameworks do not take into account the dynamic feedbacks between multiple processes of change, including the ways that responses to one process may drive or accelerate the other process, which in turn may influence the original process. These feedbacks can be channeled through physical systems, financial markets, consumer behavior, or other means. Such feedbacks affect future exposure, responses, and outcomes, hence they influence development trajectories. Many of the global environmental change discourses and frameworks acknowledge the importance of biophysical feedbacks (e.g., an ice-albedo feedback) and feedbacks across scales related to coupled human-environment systems (see Wilbanks and Kates 1999). The Millennium Ecosystem Assessment framework (Millennium Ecosystem Assessment 2003, 34) emphasizes how "humans influence, and are influenced by, ecosystems through multiple interacting pathways," and it draws attention to the linkages between both indirect and direct drivers of change. Although the indirect and direct drivers are closely associated with global environmental change and globalization processes, the framework does not explicitly consider the dynamics of these processes in relation to outcomes, but instead focuses on the consequences of ecosystem changes for human well-being. As will be shown next, the double exposure framework addresses some of these gaps, highlighting overlapping outcomes, contextual changes, and feedbacks that result from the interactions between global environmental change and globalization.

The Double Exposure Framework

The double exposure framework provides a generalized way of understanding how global environmental change and globalization interact. The framework's point of departure is that multiple global change processes are occurring both simultaneously and sequentially, creating either negative or positive outcomes for individuals, households, communities, and social groups. The framework also recognizes the role of individuals' responses and decisions as factors that both shape and are shaped by processes of global change. By emphasizing the dynamic interactions between processes, responses, and outcomes, the framework aims to elicit new insights and research questions beyond those associated with separate framings and discourses.

The presentation begins with definitions of key components and terms, including *process of global change, exposure, contextual environment, responses,* and *outcomes.* Although these terms are widely used within the various global change discourses and research frameworks, these more detailed definitions clarify how we are using the terms within the double exposure framework.

Process of Global Change

A *process of global change* entails a collective set of actions or activities that are continuously or intermittently producing large-scale transformations. Processes of global change include many different facets or manifestations, which together have coherent properties. The coherent properties of both global environmental change and globalization relate both to the scale of the processes and to the spatial and temporal connectivity between actions and outcomes occurring in different places and at different times. As was mentioned earlier, processes of global change are not new. Rather, it is the rate, magnitude, and scope of these changes that is of concern, particularly in comparison to the ability of human and ecological systems to respond to the processes. Importantly, although processes of global change entail large-scale transformations, they ultimately emanate from and are shaped by local activities, actions, and decisions.

Processes of global change may be framed collectively as "global environmental change" and "globalization," yet they also entail individual facets or components. As was discussed in chapter 1, facets of global environmental change include climate change, land use change, biodiversity loss, and stratospheric ozone depletion, among others. Each of these in turn has specific manifestations, such as higher temperatures, sea level rise, increased soil erosion, loss of pollinating species, and increased levels of ultraviolet radiation. Facets of globalization include expansion of transportation and communication networks, expansion of international trade and production, liberalization of economic policies, political integration, and cultural homogenization. Specific manifestations include changes in prices, construction of new production facilities, shifts in job opportunities and wages levels, and introduction of new consumption possibilities. The various manifestations of both global environmental change and globalization are sometimes defined as *stressors,* in that they create conditions (stresses) that represent significant deviations from the status quo and require adjustments or responses.

Processes of global change may also manifest through discrete events or *shocks* that occur unexpectedly but are of limited duration. The consequences of these events may nevertheless be long-lasting, depending on the severity of the event and the response capacity of the individual, group, or region. Examples of shocks associated with global environmental change include floods, heat waves, and forest fires. Shocks that have been linked to globalization include plant closures, national currency devaluations, and international disease outbreaks (e.g., SARS), to name a few. While many stochastic shocks are associated with normally occurring cyclical events, such as El Niño-Southern Oscillation, downturns in the business cycle, or hurricane season in the Atlantic, global change processes may make these events more frequent and more severe. Within the double exposure framework, processes of global change may manifest as either stressors or shocks.

Exposure

Exposure is the condition of being subjected to some effect or influence resulting from a process of global change. These effects may include increased temperatures, changes in exposure to ultraviolet radiation, access to external markets, and loss of state support for education. Exposure can be assessed or analyzed within a specific *exposure frame*, which is defined relative to the focus of the study. An exposure frame may represent a politically defined geographical area, such as a country or set of countries (e.g., India or the southern Africa region), or it may be defined via functional economic or ecological boundaries (e.g., an area such as the greater

New York-New Jersey metropolitan region, the Amazon rainforest, the Indus River basin, or the Arctic). The exposure frame may also represent an economic sector, such as the agricultural sector, which may be either global or confined to a specific geographic region.

The purpose of defining an exposure frame is to establish the bounds for empirical analysis, in the same way that a sampling frame sets the bounds for the statistical analysis of a larger population. Exposure thus describes how gradual transformations or discrete events occurring within a specific exposure frame directly impact upon particular units of analysis, or *exposure units*. Exposure units may be individuals, households, social groups, administrative units, communities, ecosystems, or species. Within the framework, exposure to both global environmental change and globalization varies across exposure units within any specific exposure frame (e.g., across individual households within a neighborhood, among communities within a district). For each exposure unit, exposure is a function of the magnitude and intensity of the stress or shock, as well as of the contextual conditions present within the exposure frame that make each unit of analysis more prone or sensitive to a particular change.

Contextual Environment

The *contextual environment* refers very broadly to an integrated set of conditions that influence (1) the degree or magnitude of exposure and (2) whether and how a particular exposure unit responds to global change processes. Table 3.1 provides examples of seven types of contextual factors that may influence exposure and responses: social, economic, biophysical, technological, institutional, political, and cultural. Together, these factors represent many of the key dimensions of the context in which global change occurs. Some contextual factors play a more important role than others in influencing exposure and responses, depending on the particular process. As will be discussed in the next section, individual attributes of each exposure unit also play a critical role in responses to processes of global change.

While some characteristics of the contextual environment may be homogenous across an entire exposure frame, many others will vary within the exposure frame, taking on different values for different exposure units. For example, the rate of inflation may be a general economic factor that applies to an entire exposure frame (e.g., if the exposure frame is a country such as Zimbabwe). Rates of inflation may also vary within an exposure frame (e.g., if the exposure frame is a multicountry region such as southern Africa). At any point in time, contextual factors reflect a combination of historical processes of development, such as legacies of slavery or colonialism, as well as differences in political power, which in turn affect livelihood options and entitlements (Davis 2001). Importantly, the characteristics of the contextual environment are also continuously in flux as the result of global change processes and other factors. As a result, both exposure and capacity to respond are dynamic.

Table 3.1
The contextual environment: factors that may influence exposure and responses

Type	Examples
Social	Access to health care and nutrition, educational opportunities; gender equality
Economic	Industrial structure, macroeconomic and fiscal stability, labor market characteristics
Biophysical	Climatic conditions, soil characteristics, terrain, ecosystem characteristics, water resource availability
Technological	Communication infrastructure, transport infrastructure, production technologies, water supply technologies
Institutional	Transparency of governance, government social supports, access to markets, availability of credit and insurance
Political	Freedom to act, power to influence decisions, type of social contract
Cultural	Norms and expectations, traditions, local knowledge, religious conventions

Responses

A *response* may be defined as an action taken by an individual, household, group, firm, or institution, either in anticipation of or following from exposure to some type of global change. Responses can take the form of decisions, policies, or behaviors made with the objective of influencing the outcomes of stressors and shocks associated with gradual transformations and discrete events. Responses may also be referred to as coping strategies, adjustments, or adaptations, which may minimize negative outcomes, promote positive outcomes, or limit exposure to future changes. The ability to respond to change—often referred to as response capacity, coping capacity, adaptive capacity, or entrepreneurial capacity—depends on both the contextual environment and the specific attributes of the exposure units. Wealth, education level, social capital, access to political power, and worldviews are among the many individual attributes that affect responses.

Responses can be taken in anticipation of exposure or in reaction to exposure. Anticipatory responses may include the purchase of insurance, diversification of crops or investment portfolios, improvements in infrastructure, creation of protected areas, expansion of livelihood options, and so forth. They may also include making new investments in infrastructure, education, social welfare and job training programs, or livelihood diversification. Such responses are often based on assessments of risk and strategies for social, economic, or environmental planning. Responses taken in reaction to a change may include emergency relief, sale of assets, borrowing, increased irrigation, migration from rural to urban areas, conversion of forested area to agricultural production, and closure of manufacturing plants.

Beyond contextual factors or individual attributes, it is important to recognize that responses are often influenced by actions and interventions made outside of the exposure frame, such as at other political levels. External interventions might include federal government training and education programs, humanitarian aid from international organizations, suspension of interest payments on debt by international banks, and other actions aimed at facilitating responses to change within the exposure frame. External interventions may, however, also constrain responses. For example, exchange rate adjustments in one country may reduce export earnings in another, with negative economic and social effects. Furthermore, interventions may have unintended effects if they change the contextual environment in a way that constrains future responses, such as by flooding a region with "relief" food or clothing, which may disrupt local markets and disempower producers.

Outcomes

Outcomes are the measurable or observable effects of global change processes, which can be assessed at a variety of spatial scales and from diverse perspectives. Outcomes can appear very different according to whether they are assessed from a national, regional, community, or household level or from a social group or gender perspective. For example, the net outcome of climate change on agricultural production may be negligible at the global level but substantial at the regional or local level, with very different implications for some types of farmers. Trade liberalization may be beneficial for employment growth at the national level but at the same time may create negative outcomes at the regional and local levels. In selecting and applying different outcome measures and different levels of analysis, it is important to keep in mind that outcomes are not endpoints. Rather they serve as reference points for evaluating the effects of global change processes at any particular point in time. Outcomes may also be understood as constituent parts of the dynamic contextual environment. They both influence and are influenced by contextual factors as well as by human agency, that is, the actions of individuals, leaders, and institutions (Berkes 2007).

Although outcomes can be observed, monitored, and evaluated, these acts are not value-free. What is seen as a positive outcome by some can be viewed as a negative outcome by others, depending upon the analytical lens and values of the observer. For example, a large-scale dam project may be valued by development agencies and banks because it provides electricity to rural villages and promotes participation in the global economy. However, it may be seen as a negative outcome by those who were displaced from their land or who are concerned about the environmental impacts on species and watersheds. It is important to reiterate that the particular outcomes that are prioritized and measured are likely to vary across discourses, reflecting different power relations, worldviews, and values. Some individuals, groups, governments, or corporations may prioritize economic outcomes, such as increased gross domestic product (GDP), arguing that it contributes to positive changes in human well-being. Others may prioritize biophysical outcomes, such as maximum species diversity or climate stability, arguing that they are of fundamental importance to human well-being and sustainability. Still others may emphasize social outcomes, such as access to food, health care, or jobs—or happiness and quality of life. Some examples of outcomes that may be emphasized from the perspective of the six discourses described in chapter 2 are shown in table 3.2. While some of discourses share similar outcome metrics, there is disagreement about how to achieve the outcomes. Importantly, the dominance of one discourse over another influences both the types of outcomes that are likely to be prioritized and the methods by which those outcomes are likely to be pursued.

A key point is that different outcomes and different outcome metrics may be justified or legitimated by different discourses. Often there are strong interests associated with certain outcomes, which may be backed by considerable economic, political, and cultural power (see Conca and Dabelko 2004). In the double exposure framework presented below, we will emphasize those outcomes that are consistent with human security, including those that promote equity, resilience, and sustainability. We thereby subscribe to the position of McMichael et al. (2003), which holds that readily measurable outcomes such as increased economic performance, greater energy efficiency, conservation or recreational amenities, and so forth should be considered not as ends but rather as the means to attaining desired human experiential outcomes, which include autonomy, opportunity, security, and health.

Table 3.2
Outcome measures within different global change discourses

Discourse	Outcome measure
Global environmental change	
Biophysical	Forest cover, crop productivity, species diversity, water quality
Human environmental	Livelihood diversity, ecosystem resilience
Critical	Income distribution, access to resources, gender equity
Globalization	
Benign	Economic growth, employment levels, gross national product, productivity
Malignant	Income distribution, access to resources, poverty rates, cultural diversity, human health
Transformative	Income distribution, political access, social safety nets

The Framework

The double exposure framework helps to structure thinking and analysis of the interactions between processes and outcomes of global change. In the framework, global environmental change and globalization manifest as gradual or sudden changes that differentially affect exposure units such as individuals, communities, social groups, or regions, resulting in measurable outcomes. Within the framework, *outcomes* depend on both *exposure* to each global process and *responses* taken by the exposure unit. As was described above, exposure to each *process* is influenced by the characteristics of the change (e.g., direction, rate, magnitude, intensity, and spatial extent) and by factors in the *contextual environment* (e.g., social, economic, biophysical, technological, institutional, political, and cultural conditions). Responses, which may include any actions taken either in anticipation of or following from exposure, are conditioned by factors in the contextual environment and by the individual attributes of each exposure unit (e.g., wealth, political access, social capital).

Figure 3.1 illustrates the various components of the framework from the perspective of a single exposure unit. Processes of global environmental change and globalization are represented as partially overlapping triangles. These processes manifest within a specific exposure unit, portrayed as an oval. The extent or magnitude of exposure to the processes is depicted as the intersection between the triangles and the oval. Responses to the processes are represented by an arrow leading from the exposure unit to a square representing outcomes. Outcomes are depicted within the figure as separate from the contextual environment to emphasize that an outcome reflects measurable conditions at a specific point in time.

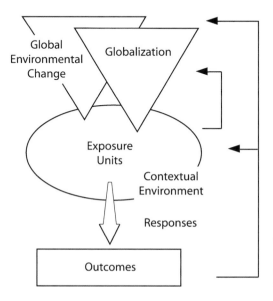

Figure 3.1
A schematic diagram of the double exposure framework.

The framework incorporates dynamic linkages between the components. Processes may alter the contextual environment, responses may affect the processes, outcomes may affect responses, and so forth. Transformations to the contextual environment can be represented in figure 3.1 in two ways:

1. The oval surrounding the exposure unit may change shape, signifying a direct transformation of the context as the result of either or both processes of global change. Examples of contextual transformations associated with global change processes include changes in biophysical conditions, such as melting of permafrost in the Arctic as the result of climate change, and changes in institutional supports as the result of new international agricultural trade agreements.
2. The arrow leading from responses and outcomes to the contextual environment in figure 3.1 signifies transformations of the context that may follow from outcomes of either process. Examples of such responses may include infrastructure development, such as expansion of sea walls in response to the threat of sea level rise, or they may include regulatory reforms such as adoption of new zoning rules to limit construction in a flood-prone region.

Dynamics are also incorporated in the framework through recognition that processes and outcomes are often reflexive. Within figure 3.1 these types of circular linkages, which we term *feedbacks*, are depicted by the arrows leading from the contextual environment and from responses and outcomes back to the process triangles.

Within the framework, outcomes describe the consequences of exposure to global change for individual exposure units. It is important to recognize that outcomes are highly variable across exposure units as the result of differences in exposure, contextual factors (including individual attributes), and responses. As will be shown in chapter 5, exposure units that are negatively affected by exposure to one process are often similarly affected by exposure to the other process. At the same time, those exposure units that are positively affected by one process tend to be positively affected by the other process. These overlapping positive and negative effects help to explain how and why the exposure to the two processes may contribute to growing social and economic inequalities and polarization.

In a world of growing connectivities, outcomes are also increasingly linked across both space and time. Figure 3.1 focuses on a single exposure unit, but it is important to note that outcomes and responses that occur within one exposure frame may have widespread influences on other exposure frames as well. These types of spatial linkages will be explored further in the discussion of Hurricane Katrina in chapter 6. Temporal linkages become apparent when we recognize that many types of changes to the contextual environment, particularly those associated with global environmental change, are not easily reversible and will thus influence the exposure and response capacity of future generations. These types of linkages over time will be illustrated in chapter 7 in the investigation of Arctic melting.

———— *Conclusion*

Frameworks set boundaries around issues—boundaries that influence the questions that are asked, the research approaches that are prioritized, and the responses and policies that are pursued to address processes and their outcomes. The double exposure framework provides a tool for examining multiple types of interactions between two transformative global processes. In particular, the framework illuminates the distributional effects of both processes over different scales and reveals negative and positive feedbacks and synergies between the two processes. The framework can also motivate new questions about the way that processes, outcomes, and responses interact, as well as what these interactions mean for human security. Although different methodologies and levels of analysis may be emphasized by different research communities, the framework provides a generalized way to explore the dynamics of global change.

The next chapter expands upon the double exposure framework to identify and describe three pathways of interaction between the two processes. These three pathways—each of which is explored in detail in later chapters—help to articulate some of the direct and indirect ways that global environmental change and globalization together create uneven outcomes, influence vulnerability to shocks and stresses, and contribute to accelerating rates of social and environmental change. How these interacting processes and their outcomes are addressed, both today and in the coming years, will have a decisive influence on human security in the twenty-first century.

Pathways of Double Exposure

4

All paths have branching points: some go off in unexpected
directions, others into dead-ends. This means that we need both a
good map and a clear sense of the direction we wish to travel.
—Peter Dicken, *Global Shift: Mapping the Changing
Contours of the World Economy*

The double exposure framework draws attention to the interactions between two
transformative processes of change. It emphasizes that global environmental change
and globalization, which are often treated in contemporary discourses as though
they are distinct and unrelated, are in reality closely connected. The framework
also provides a conceptual tool for investigating how the two processes together
influence present and future outcome trajectories. In this chapter we further elabo-
rate on the framework through a description of some of the possible pathways of
double exposure. The pathways, which we term *outcome double exposure, context
double exposure,* and *feedback double exposure,* describe how the processes interact
over time and across space.

By describing several pathways of double exposure, we are able to articulate
and explore specific interactions between global environmental change and globalization, and to consider the implications for equity, resilience, and sustainability.
The pathway of *outcome double exposure* shows how overlapping exposure to the two
processes can exacerbate regional and social inequalities. The pathway of *context
double exposure* shows how the processes, both separately and together, can increase
the vulnerability of individuals, communities, and social groups to shocks and
stresses of all types. The pathway of *feedback double exposure* emphasizes how global
change processes can generate responses that can amplify the processes, leading to
new cycles of double exposure. While each pathway emphasizes a different type of

42

interaction between the processes, it is important to recognize that the pathways may occur both simultaneously and consecutively. Moreover, it is seldom one pathway alone that defines outcomes.

(handwritten margin notes: "why pathways?" "For All component" "confusing" "Pathways to what?")

—— *Outcome Double Exposure*

Outcome double exposure stresses the direct and immediate outcomes of global change processes. This pathway, illustrated in figure 4.1, highlights situations in which the same units are exposed to the effects of both globalization and global environmental change. Under outcome double exposure, different exposure units (e.g., individuals, households, communities, or social groups) located within an exposure frame (e.g., a region, country, or sector) are exposed to discrete events and/or gradual changes emanating from both global environmental change *and* globalization. Exposure to each process varies across exposure units, depending upon the magnitude, extent, and rate of the changes and on conditions within the contextual environment. Exposure to both processes often occurs simultaneously; thus outcomes may not be attributed to one particular process by any exposure unit. Nevertheless, the outcomes are often amplified by the two interacting processes, a situation that further transforms the contextual environment and influences exposure, responses, and outcomes of future change. The consequences of outcome double exposure thus can influence the direction of development trajectories. [2]

Although global environmental change and globalization processes may not always be directly related, outcome double exposure explains why those who are

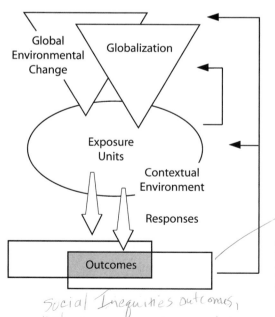

Figure 4.1
The pathway of outcome double exposure.

(handwritten notes around figure: "Rectangles will completely NEVER develop inequity! Always rather Have a not so broad context that is completely separate below" "Social Inequities outcomes, not biophysical — social... Can we make it that?")

negatively affected by one process may also be negatively affected by the other. This situation occurs when the same (or similar) contextual conditions influence exposure and the capacity to respond to both processes, and hence outcomes. For example, Eriksen and Silva (2008) show that limited access to markets in Mozambique makes it difficult for some households to sell their products *and* to acquire inputs such as drought-tolerant seeds to prepare for climatic stresses (Eriksen and Silva 2008). For these households, contextual conditions may contribute to similar outcomes from increasing market liberalization and climate extremes.

This co-occurrence of negative outcomes is not trivial. In some cases the outcomes may be not simply additive but synergistic, pushing some who are affected beyond their limits in terms of what they can cope with or respond to. For example, in a study of Mexican farmers, Eakin (2003, 2006) showed that globalization-related market uncertainty and price volatility can interact with climatic risks to undermine livelihood strategies. In the case of irrigated vegetable farmers in Puebla, Mexico, she found synergies in the effects of climate and market risk not only in terms of their direct impacts on harvests but also in national production and price trends (Eakin 2003). For example, when farmers in high-volume vegetable-producing regions of Mexico could not export their produce to the United States, they turned to domestic markets, thereby driving down prices for producers of lower-quality products, who operate with smaller volumes (Eakin 2006).

Often the same individuals, communities, or groups that benefit from global environmental change also benefit from globalization, at least when viewing "snapshots" of outcome double exposure. The benefits or positive outcomes from global environmental change are not restricted to positive changes in the biophysical environment. Drawing again on the work of Eakin (2003, 2006) in Mexico, we can see that she finds that differential patterns of climate variability affect crop prices and present challenges for some farmers but opportunities for others. When damaging droughts occurred in the Mexican states of Sinaloa and San Luis Potosi, farmers in Puebla were able to take advantage of higher prices for husk tomatoes and cabbage in domestic markets.

Outcome double exposure raises important questions related to equity. Equity is considered an important prerequisite for both social development and human security (UNDP 2005, 2007; UN 2005a; World Bank 2006b). The concept of equity is associated with the freedom from bias or favoritism and entails outcomes that are perceived as fair to all concerned (O'Brien and Leichenko 2006). The idea of equity is often related to questions of justice, including the notion that there should be equal treatment for equal cases (Rawls 1972; Boulding 1978; Smith 1994; Ikeme 2003). Equity also has a temporal dimension, in that outcomes for future generations should be treated with the same consideration as outcomes in the present (O'Brien and Leichenko 2006). Although there is little consensus over the causes and remedies of inequities among different global change discourses (Koivusalo 2006), recent economic research shows that "ceteris paribus, the more egalitarian a society, the better its growth record and growth potential" (Sanchez 2003, 1988). Moreover, relative income equality is also associated with greater social cohesiveness and a higher quality of democratic governance (Sanchez 2003, UN 2005a).

One important equity-related question raised by outcome double exposure is: Who is most likely to experience negative outcomes related to both processes? In many cases, it is those whose contextual environment maximizes exposure and minimizes the capacity to respond. Poverty, which itself can be considered both a process and an outcome, is often closely associated with a series of negative outcomes that lead to marginalization (Ehrenreich 2001; CPRC 2004; Glasmeier 2005). Indeed, concentrated poverty among specific groups is often a function of political and economic power, including control over resources and working conditions (Lister 2004). Responding to global change from a point of marginalization may be difficult, particularly if one is highly exposed to stressors and shocks associated with both global environmental change and globalization. In contrast, those who are most able to take advantage of the opportunities associated with both processes are those who live within a context where biophysical, social, economic, cultural, technological, institutional, and political conditions either minimize exposure or maximize the capacity to respond (Roberts and Parks 2006).

The following vignette of outcome double exposure shows how contemporary initiatives to marketize water resources coincide with changes in global water systems, resulting in overlapping negative outcomes for individuals, households, and communities that have limited access and entitlements to water.

Water Marketization and Equity in South Africa

The marketization of water is one aspect of globalization that has been widely discussed from diverse perspectives within the three globalization discourses presented in chapter 2. Water marketization, considered by Conca (2006) as the process of creating the economic and policy infrastructure for treating water as a marketed commodity, refers to a "a broader set of linked transformations related to prices, property rights, and the boundary between the public and private spheres" (Conca 2006, 216). This process has been driven by neoliberal economic reforms, including structural adjustments, trade liberalization, and privatization (Barlow and Clarke 2002; Conca 2006; UNDP 2006). Marketization can take different forms, ranging from a complete transfer of state assets to private corporations to different models of public-private partnerships. Among the latter, schemes can include delegation of operational and managerial functions to private companies (while infrastructure and equipment remain in public hands) or the transfer of service responsibilities to individuals, communities, and nongovernmental organizations (McDonald and Ruiters, 2005).

Debates over water marketization reflect different valuations of outcomes (PPIAF 2002; WaterAid and Tearfund 2003). Proponents of water marketization and privatization argue that it promotes efficient and reliable use and distribution of water, with a focus on service and performance (World Bank 2006a). Opponents counter that marketization transforms water from a human right into a commodity and that it favors profit maximization and increased consumption over universal access to water and sustainable use (Barlow and Clarke 2002). Under marketization schemes, the poorest people in poor countries tend to pay more for their water than the rich, who often receive municipal water subsidized by governments (Barlow and Clarke 2002). Reliance on private vendors who charge exorbitant prices, or on distant and/or polluted water sources, adds to the cost of water for those not connected to public systems (Devas 2002). As García-Acevedo and Ingram (2004, 20) observe: "The struggles among water users within nations and across borders have left a string of winners and losers. Water serves as a kind of tracer element, identifying those who wield political and economic power and those who don't."

Water marketization efforts worldwide have profound implications for individuals and communities that depend on free and open access for survival (Barlow and Clarke 2002; Gleick et al. 2002; WaterAid and Tearfund 2003). They also have potentially enormous consequences for the world's rivers, watersheds, and freshwater supplies (Conca 2006). More important, the impacts of water marketization coincide with ongoing changes in the global water cycle as the result of global environmental change. Of primary concern are the availability and quality of freshwater supplies to both sustain life and maintain freshwater ecosystems. Freshwater supplies worldwide are threatened by overpumping of groundwater, pollution from point and nonpoint sources, and diversion of water for irrigation and hydropower uses (Vörösmarty et al. 2004). As was discussed in chapter 1, these changes are reflective of the types of cumulative changes that occur on a local level yet affect a significant share of total global resources. Not surprisingly, many of the most alarming facets of global climate change concern threats to freshwater resources as the result of sea level rise, melting of glaciers, and increased frequency and magnitude of drought and floods (McCarthy et al. 2001; IPCC 2007a).

Outcomes from globalization and global environmental change are both overlapping and interrelated with respect to water resources. For example, many of the pressures on local water supplies are connected both to globalization-related changes, such as expansion of irrigation schemes for water-intensive agricultural production of export crops, and to changes in water availability due to climate variability (O'Brien et al. 2004). In the case of South Africa, where climate variability and long-term climate change are of increasing concern (Hulme 1996; O'Brien and Vogel 2003; Vogel 2005), a number of cities and towns have recently privatized water supplies (McDonald and Ruiters 2005; Funke et al. 2007) . The marketization of water means that water is no longer supplied as a free or nominally priced public good but instead follows the rules of the market. When water supply decreases, as during years of drought, water prices can potentially increase to ensure full recovery cost for infrastructure and service delivery. As McInnes (2005) notes, to be sustainable, a corporatized water service provider must maintain a surplus, even if doing so means restricting or cutting services to enforce payment from households.

The South African constitutional reform of 1996–1997 recognized socioeconomic rights, including guaranteed access to sufficient food and water (Jones and Stokke 2005, Conca 2006). The National Water Act of 1998, a key milestone for water reform in South Africa, emphasized principles of decentralization, equitable access, efficiency, and sustainability (Funke et al. 2007). Nonetheless, the contextual environment in South Africa is characterized by deeply entrenched social and economic inequality (a legacy of apartheid) and limited political representation for many groups, and there are numerous challenges to the implementation of the National Water Act (Funke et al. 2007). There is, among other things, a tension between aggressive cost recovery and service cutoff practices and the language of rights and social equity (McInnes 2005). Those who have been most negatively affected by water privatization are also significantly affected by lack of available water supplies as the result of overpumping, pollution, and drought. Many poor residents of urban townships in particular have been double exposed to both increased prices and reduced water availability. Wealthy residents, in contrast, benefit from improved water supply and quality associated with marketization, and they remain relatively unaffected by changes in availability. The inequities may become more pronounced in the future, as surface water supplies are projected to decrease with climate change, particularly in the Western Cape region of South Africa (De Wit and Stankiewicz 2006).

Water debates within global environmental change and globalization discourses generally do not consider how individuals and communities are likely to be affected by *both* processes. Viewing them through separate discourses, one sees an incomplete picture of how individuals, households, or communities are affected by changes in water access and availability. The efficiency and service gains from privatization may not, for example, accrue to those who are most vulnerable to water stress associated with climate variability. The pathway of outcome double exposure demonstrates how these two processes may overlap within a particular exposure frame to exacerbate existing inequities. In chapter 5 we will provide a more detailed example of outcome double exposure, showing how two specific manifestations of global environmental change and globalization, namely climate change and agricultural trade liberalization, differentially affect districts, farmers, and social groups in India.

Context Double Exposure

Context double exposure stresses how new conditions associated with both global environmental change and globalization may change the contextual environment and increase vulnerability to shocks and stresses of all types. Vulnerability has been defined as susceptibility to negative outcomes, with the most vulnerable being those who are most exposed to perturbations, who possess a limited capacity for adaptation, and who are least able to recover (Bohle et al. 1994). As is illustrated in figure 4.2, context double exposure involves changes to the contextual environment (i.e., the shape of the oval), which affect not only exposure (i.e., the overlap with process triangles) but also response capacity (i.e., the width of the response arrow).

In many cases, exposure to global change processes directly alters the contextual environment, thus affecting future exposure, responses, and outcomes. Such was the case when Hurricane Mitch struck the city of Tegucigalpa, Honduras. Changes to the biophysical and social context included deforestation of the hill slopes around the city and the expansion of the population of rural in-migrants

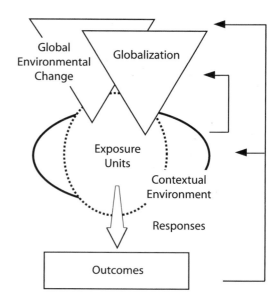

Figure 4.2
The pathway of context double
exposure.

living in these marginal areas (Segnestam et al. 2006). The capacity for responding at the national level was influenced by structural adjustment programs,
which emphasized privatization, modernization, and increased efficiency in the
public sector, and by strengthening of the financial system. This combination of
changing biophysical, social, economic, and institutional conditions made some
individuals and communities of Tegucigalpa highly vulnerable to hurricanes and
other flood hazards (Boyer and Pell 1999; Girot 2002; Pelling 2003a; Segnestam
et al. 2006). When Hurricane Mitch passed through Central America in 1998, it
hit upon this context, creating an enormous disaster, with over 6,000 lives lost in
Tegucigalpa alone.

Changes in the contextual environment may also result from local responses
and adaptations to processes of global change. Research by Neil Adger (1999;
2000a) in Vietnam shows that coastal flooding as the result of typhoon-related
storm surges is a continual threat along the country's 3,000-km coastline. Because
coastal flooding is a "normal" risk in Vietnam, various institutions have evolved to
help reduce risk and enhance response capacity. Under globalization, some facets
of the Vietnamese institutional context have changed dramatically. In particular,
formal state institutions have undergone a series of neoliberal economic and political shifts since the early 1990s that signaled a transition away from a centralized,
state-planned economy toward a market-oriented economy. These changes, which
included reform of land tenure practices toward private ownership of land and
reductions in public expenditure for flood control, altered the formal institutional
context affecting both exposure and responses to coastal flood hazards. At the same
time, however, other facets of the institutional context have been much slower to
change, particularly worldviews that are influenced by prior communist history

and cultural norms. Consequently, there was an expectation that authorities would take responsibility for protecting communities from flood hazards. This combination of rapid, globalization-related institutional changes and lack of change in worldviews is compromising resiliency to both current "normal" flood hazards and future climate change-related events (Adger 1999, 2000a).

While context double exposure emphasizes exposure and responses to global change, it also recognizes that contextual changes associated with both processes may exacerbate the effects of many other types of shocks and extreme events. The 2004 Asian tsunami, for example, was the result of a naturally occurring geological event, yet the devastation in countries such as Sri Lanka and Thailand was exacerbated by contextual changes associated with both global environmental change and globalization, particularly the loss of mangroves due to expansion of shrimp farming for global markets (Adger et al. 2005; EJF 2006).

Over time, the contextual changes associated with both processes may influence the resilience of social-ecological systems. Indeed, it is becoming clear that resilience is more tightly linked to global processes today than in the past (Adger et al. 2005). More than the capacity to absorb shocks, resilience is considered a key factor influencing adaptive capacity (Klein et al. 2003). Resilience "concerns the capacity for renewal, re-organization and development, which . . . is essential for the sustainability discourse" (Folke 2006, 253). The following vignette illustrates context double exposure, showing how neoliberal changes in social welfare policies and institutions influenced the capacity to respond to the 2003 Paris heat wave.

Social Resilience and the Paris Heat Wave

During the summer of 2003 a severe heat wave affected most of Western Europe, causing more than 22,000 heat-related human deaths, nearly 15,000 of which occurred in France (Pirard et al. 2005). It also resulted in the deaths of thousands of farm animals and a dramatic decline in agricultural production. During the heat wave, average summer temperatures (June, July, August) in many parts of Europe were approximately 3°C above normal (Schär et al. 2004), with a particularly severe peak in the first two weeks of August. Though the heat wave cannot be decisively attributed to climate change (Chase et al. 2006), several studies have demonstrated that this type of heat event is statistically unlikely to have occurred randomly, without human-induced climate change (Schär at al 2004; Stott et al. 2004). In any case, this type of event is symptomatic of the types of extreme climate events that are likely to recur more frequently—and more intensely—with increases in atmospheric concentrations of greenhouse gases (Meehl and Tebaldi 2004; Beniston 2004).

High temperatures played an important role in the disaster that ensued in France. However, as Lagadec (2004) points out, heat was not at the core of the 2003 fiasco. Rather, organizations and people did not have the intellectual and practical frameworks to respond to the heat episode (Lagadec 2004). Among the people who died in France, most were elderly and female, and most lived in and around Paris (Pirard et al. 2005). Both individual attributes and contextual factors played a decisive role in determining who survived the heat wave and who did not (Poumadère et al. 2005). Many factors that were identified by Klinenberg (2002) in regard to the 1995 Chicago heat wave also influenced individual outcomes in Paris, including the level of social isolation, socioeconomic status, and age and health conditions. The conditions of the built environment—for example, the type of housing, access to air-conditioning, and presence or lack of parks, green spaces, and shade trees—were also important influences of outcomes.

The institutional context in France has undergone changes associated with neoliberal policies and New Public Management (Ferlie et al. 1996). The French administrative system has traditionally been centralized and hierarchical, intervening in all aspects of socioeconomic life. As Cole and Jones (2005, 569) point out, "[s]uch state interventionism (*dirigisme*) has declined since the 1980s, a by-product of the globalization of world markets, deregulation, and the rise of consumerism." The shift toward more individualized, consumer-oriented approaches to social services, whereby people receive such services only when they ask for them, was also documented by Klinenberg (2002) in the Chicago case. The New Public Management agenda, which includes "a stress on greater discipline and parsimony in resource use in order to encourage public-sector bodies to maximize their use of dwindling public resources" (Cole and Jones 2005, 568), may decrease the capacity to respond to extreme weather events.

The social context has also undergone transformations in France. Family networks in particular have changed, such that the elderly are increasingly living separately from younger generations. This phenomenon, which is growing in almost every advanced economy, is rooted in broad-based societal and demographic shifts, many of which would not be considered negative by other criteria. In the case of France, because of changes in the nature of labor markets, increased mobility, and increased wealth, many younger residents left Paris during the heat wave for vacation homes. As a result, less mobile elderly residents were not able to rely on family members to help them manage the heat. The state retreat under New Public Management thus coincides with a growing need for state intervention to help those who might have relied on family during a different era.

All these changes in the institutional and social context were, however, only part of the picture. Because the heat event was outside the boundaries of what people had experienced in climate extremes, many were not knowledgeable about how best to respond (Keatinge 2003). Although the heat wave temperatures were well within the range of normal for cities in some other parts of the world, such as South Asia, they far exceeded what people were accustomed to in Western Europe. Most of the resources that were needed to combat the heat were readily available in France: ice and water, covering windows to keep out the sun, freezing clothes before wearing them, placing humid towels next to a fan (Lagadec 2004). Under a different set of expectations and experiences, the Paris heat wave could have been little more than an annoyance involving a short-term disruption of routine activities; instead, it was transformed into a disaster.

Heat waves represent the number one major risk among "natural hazards" in postindustrial societies (Poumadère et al. 2005). Yet the negative outcomes of the

2003 heat wave have prompted some positive responses. In particular, they have inspired other French and European cities to reexamine their efforts to plan for heat waves; indeed, they led the French government to draw up a "Plan Canicule 2006" to review and revise disaster response plans.[1] Through the global media, exposure and negative outcomes in one location revealed problems within both the contextual environment (i.e., lack of institutional capacity) and cultural expectations about climate-related events. This knowledge may, in turn, lead to actions that change the institutional context and create social awareness about climate extremes in other locations, as well as drawing attention to the vulnerability of the elderly. The double exposure framework recognizes the connections across locations and scales as part of the transformation of contextual environments. It emphasizes a dynamic web of interactions over time and across space, whereby future outcomes in any location may be altered as the result of processes and outcomes in other locations.

[handwritten margin note: A positive response!]

[handwritten margin note: This is becoming increasingly in c the rise of both g.e.c & globalization]

Feedback Double Exposure

The pathway of feedback double exposure highlights temporal linkages between global change processes, outcomes, and responses. This pathway emphasizes the dynamics inherent within the double exposure framework, showing how actions taken in response to either or both processes may contribute to the drivers of global change. A feedback is defined here as a return to the input of a part of the output of a system or a process. As is illustrated in figure 4.3, the feedbacks to both processes may result from responses to either contextual changes or outcomes. In some cases these feedbacks may accelerate either or both processes of global change, potentially *[handwritten: or decelerate]*

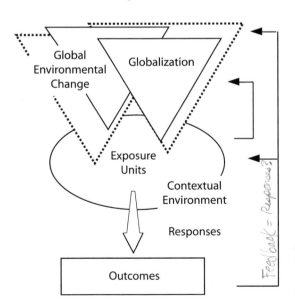

[handwritten note along figure: Feedback = Responses?]

Figure 4.3
The pathway of feedback double exposure.

increasing future exposure through changes in the magnitude, frequency, and extent of shocks and stressors. These can result in an iterative cycle, whereby responses in the present may inadvertently promote global change processes in the future, leading to new rounds of double exposure. Alternatively, the linkages between the two processes may have a dampening effect, such that responses to one process may ameliorate the drivers of the other process.

The potential for feedbacks to either amplify or dampen the processes may be illustrated via the example of tropical deforestation (Rudel 2002; 2005; Hecht 2005). In the case of the Brazilian Amazon, linkages between soybean production for export and destruction of primary forests are well documented (Fearnside 2001; Laurance et al. 2001; Hecht 2005; Nepstad et al. 2006). Land clearance for soybean production also creates feedbacks that perpetuate deforestation in other regions. These feedbacks occur as the result of infrastructure investments, including highway construction for truck transport and waterway development for barge transport, both of which facilitate the transport of soybeans to international ports. These investments have increased accessibility of vast tracks of the Brazilian Amazon for extraction of timber and other commodities, which can be sold in global markets. The investments have also opened new areas for settlement, leading to further clearance of forests for crop production (Fearnside 2001; Soares-Filho et al. 2006).

In other forest regions, however, globalization processes have had positive effects, promoting forest regrowth (Rudel 2002, 2005; Nepstad et al. 2006; Hecht et al. 2006). Hecht et al.'s (2006) research in El Salvador, for example, shows how agricultural contraction due to market factors such as declining commodity prices, capital scarcity, and stagnant agricultural wages, along with male out-migration and increased remittances from abroad, have transformed rural areas from sources of crop production to suppliers of environmental services. Their work shows how a range of changes linked to globalization have enhanced forest resurgence, demonstrating the complexity of forest trends in globalized economies (Hecht et al. 2006).[2]

Feedbacks may also be illustrated via examination of the linkages between globalization and land use changes in the Philippines. In response to a loss of livelihood options resulting from the expansion of industrialized agriculture, many rural women in the Philippines migrate for contract work overseas. Their actions not only contribute to processes of transnational migration, but also create feedbacks which promote further land use change within the Philippines. McCay (2005), who provides a gendered interpretation of landscape change in the Philippine province of Ifugao, shows that the remittances from women overseas are often used by solo fathers to convert subsistence rice paddies to market-oriented production of green beans. The proliferation of these cash-crop gardens—which are seen to reflect the predomination of a "male" lens on the landscape—is having substantial impacts on both soil quality and water availability, and it may potentially undermine the sustainability of rice production in the region (McCay 2005).

The pathway of feedback double exposure raises important questions about the sustainability of current development trajectories. In particular, it draws attention to the ways that global environmental change and globalization may interact to undermine sustainable development, a concept that emphasizes the idea of using natural resources in a manner that meets the needs of the present generation without compromising the ability of future generations to meet their needs (WCED 1987; Guimarães 2004).[3] Although definitions of sustainability vary, the broader goals are to foster economic and social practices that contribute to the maintenance of environmental quality, diversity of species, and preservation of ecological functions for the benefit of both humans and other living things (Folke et al. 2002; Kemp and Parto 2005). Feedback double exposure thus raises issues of intergenerational equity, social justice, and transfrontier responsibility, all of which are closely connected to sustainability concerns (Gleeson and Low 2000; Agyeman et al. 2003; Swart et al. 2003).

Feedback double exposure also shows that the relationship between processes and responses is not a straightforward, linear one. The processes are often intertwined, such that the "solution" to one problem may drive other processes, creating both intended and unintended outcomes. When linkages between processes are ignored, as often occurs when global environmental change and globalization are addressed as separate issues, there is potential to perpetuate outcome trajectories that are unsustainable over time and space. The sustainability of outcome trajectories is nevertheless closely linked to the types of outcomes that are prioritized. As John Robinson (2004, 383) puts it, sustainability can be considered "the emergent property of a conversation about what kind of world we collectively want to live in now and in the future." In the following vignette, the example of suburbanization in Beijing, China, illustrates how linkages between the two processes create feedbacks that contribute to global environmental change.

Suburbanization and Sustainability in Beijing

One of the major new forms of land use change in urban areas throughout the world entails construction of low-density, suburban housing developments that are spatially removed from urban core areas. This development pattern, which has been prevalent in the United States for more than 50 years, is becoming increasingly common in cities throughout the world (Leichenko and Solecki 2005; Lankao 2007). Elements of this changing urban form include expansion of the spatial extent of cities, reductions in urban density as cities move toward more decentralized, multinodal, forms, and homogenization of the housing consumption preferences of the middle- and higher-income populations for exclusive, spatially segregated, suburban-style communities (Solecki and Leichenko 2006).[4]

Feedback double exposure is exemplified in the case of Beijing, a city with a population of over 13 million. There has been a recent boom there in new construction of suburban housing developments containing "American-style" townhouses and single-family homes with attached garages (Wu 2004). These developments are located primarily on the outskirts of the city and are spatially segregated from the poorer, rural-migrant communities that are also found throughout Beijing's urban fringe (Gu and Shen 2003). In many cases the new suburban communities have Western names such as Yosemite, Orange County, or Vancouver Forests, and some of them are designed as direct replicas of similar projects abroad. In the case of Orange County in Beijing, the blueprints and designs are drawn from a U.S. community of the same name in California (Wu 2004). Although many of these Western-style housing communities serve as gated communities for foreigners living in China, this type of development is now becoming increasingly popular for middle- and upper income Chinese (Wu and Webber 2004). Orange County, for example, is entirely populated by Chinese homeowners (Wu 2004).

Global change processes have played a critical role in the spread of new suburban developments in Beijing. In the 1980s, globalization-related institutional changes opened the Chinese economy to international trade and foreign direct investment, and to participation in international financial markets. For Beijing, these neoliberal changes led to tremendous growth in foreign direct investment and an influx of well-paid foreign workers as multinational corporations have sought to place regional headquarters in China's capital city (Wu and Webber 2004). This influx of new investment dovetailed with further neoliberal institutional changes during the 1990s which were intended to privatize housing markets, initiate housing finance mechanisms, and shift control over land development and planning from the government to the private sector (X. Zhang 2000; T. Zhang 2000; Wang 2001; Wu and Webber 2004).

This suite of neoliberal institutional reforms during the 1980s and 1990s changed contextual conditions and facilitated exposure to other global change processes within Beijing, including the proliferation of an ideology of consumption. In particular, the influx of foreign investment has contributed to a growing class of wealthy transnational consumers. These new consumers—many of whom are transplants from the United States, Canada, and Europe—have tended to embrace a global ideology of consumption and with it the desire for higher-end, Western-style housing in suburban communities as well as a better quality of life and access to environmental amenities (Sklair 2002; Leichenko and Solecki 2005). At the same time, because of the growing openness of the Chinese economy to media and print advertising, these consumption desires are not limited to higher-income residents: consumers at all levels in Beijing are exposed to television images and billboards depicting all types of consumer products from cell phones to new suburban houses. Fulfilling these new consumption desires is also increasingly possible as large retail outlets such as Walmart and Home Depot begin to appear in Beijing and other large Chinese cities.

In addition to new consumption preferences, neoliberal policies have also been critical in enabling the land use changes associated with suburban development, including the conversion of agricultural to urban land use. In Beijing, the amount of urban land (as opposed to arable land) in the peripheral areas around the city doubled during the 1990s, from 416 km^2 in 1990 to 841 km^2 in 2000 (Tan et al. 2005). Policy changes that have enabled this rapid conversion from agricultural to urban uses include devolution of central control over land use planning, development of land markets and changes in laws to allow governments to lease land

(T. Zhang 2000, Tan et al. 2005). The proliferation of exclusive, spatially segregated, suburban-style communities represents only a portion of this new urban land—other uses include industrial parks, roads, and high-density residential apartment construction. Nonetheless, the new suburban communities play a critical role in fostering responses that lead to further land use change, including the construction of new shopping centers, parks, and highways to serve the consumption and amenity needs of these residents (Leichenko and Solecki 2008).

The construction of low-density suburban developments also leads to behaviors and activities that contribute to other processes of global environmental change, including more rapid buildup of greenhouse gases in the atmosphere. Residents of low-density, suburban communities use more energy per capita than do residents of high-density, inner-city areas, because of larger house size, higher levels of material consumption for furnishings and electrical appliances, and greater reliance on automobile transport (Leichenko and Solecki 2005; Neff 2007; Lankao 2007). Within large Chinese cities such as Beijing, the expansion of automobile usage due to suburbanization is increasingly recognized as a major source of greenhouse gas emissions, particularly carbon dioxide (Gan 2003; Lee 2007). Automobiles are also causing significant air pollution from carbon monoxide, sulfur dioxide, and other pollutants (Gan 2003). While these local pollutants are most commonly associated with smog and other urban air quality problems, they are also beginning to be recognized as contributors to climate change (Rypdal et al. 2005).

Feedback double exposure stresses the dynamic linkages between globalization and global environmental change, showing how contextual changes, outcomes, and responses from one process not only may perpetuate the other process, but also may exacerbate the initial process. In the case of Beijing, we have stressed the negative environmental feedback between land use change and globalization, which may undermine sustainability. Yet it is important to recognize that there is also the possibility of positive linkages between processes of global change, whereby, for example, increasing globalization may promote new environmental reforms. In Beijing, international actors have introduced more energy-efficient "green" buildings and water recycling systems, and so globalization may eventually lead to homogenization of building and environmental standards and thus to an improvement in local environmental conditions (Presas 2004). As will be discussed later, these types of positive connections between the two processes provide opportunities to address global change.

Conclusion

The double exposure framework helps to unravel the complexity of interactions between two transformative global processes. Each of the three pathways described in this chapter highlights a different type of interaction between the processes, and each draws attention to important questions about human security. Outcome double exposure raises questions of equity, which inevitably lead to questions of power and politics. Is growing polarization desirable or just? Are outcomes that benefit some at the expense of others fair? Are the winners from either process responsible for compensating the losers? Context double exposure raises questions of vulnerability, as seen from a dynamic context. Is the capacity of individuals and households to respond to change being undermined by the processes themselves? Are contextual changes compromising the resilience of social and ecological systems? Feedback double exposure forces us to confront issues of sustainability. Are the positive and negative outcomes that are observed at any instant sustainable over time and across spatial scales? Will an accelerating pace of change as the result of both processes ultimately exceed the response capacity of ecological and social systems? While each of the three pathways emphasizes challenges to human security posed by the interactions between the two processes, it is important to recognize that these interactions may also be used to identify openings that can enhance human security. Processes can be changed, contextual conditions can improve, and interventions can influence outcomes and development trajectories. To create these positive synergies, there is a need—first and foremost—to consider the two interacting processes together.

Each of the next three chapters will explore these different pathways of double exposure in more detail. In chapter 5 we will investigate outcome double exposure via exploration of the agricultural sector, showing how the two processes may interact to create increasing inequalities for rural producers. In chapter 6 we will examine context double exposure and show how the interactions between the two processes within cities are creating greater vulnerability to hazards and extreme events. In chapter 7 we will draw attention to feedback double exposure to illustrate how the interactions between two processes contribute to accelerating rates of change with the Arctic region.

Uneven Outcomes and Growing Inequalities

5

In spite of the compelling case for redressing inequality, economic and noneconomic inequality has actually increased in many parts of the world, and many forms of inequality have become more complex and profound in recent decades.
—UN, *The Inequality Predicament: Report on the World Social Situation 2005*

Global environmental change and globalization each creates uneven outcomes across regions, sectors, and social groups. Often described in terms of "winners and losers," these differential outcomes are especially evident in the agricultural sector, where there is growing polarization between large and small farmers, landowners and landless laborers, and productive and marginal regions. These growing asymmetries are also evident within agricultural commodity chains, where wealth and control over production are increasingly concentrated among large supermarket chains and agribusiness firms (Goodman and Watts 1997; Jodha 2000; Devereux and Maxwell 2001; Conca 2002; Guy Robinson 2004; Silva 2007). While some farmers, regions, and firms are benefiting from both global processes, many others are experiencing great difficulty in coping with and adapting to the new conditions. Furthermore, the overlapping outcomes of the two processes often have compounding effects that perpetuate or exacerbate inequalities over time.

This chapter draws on the pathway of outcome double exposure to explore the uneven consequences of global environmental change and globalization on agriculture and rural livelihoods. First the consequences of two of the most far-reaching and transformative types of global change are examined, namely climate change and trade liberalization. Although the various literatures on these topics prioritize different outcome metrics, there is widespread agreement that both processes are creating winners and losers. The question is, do the winners and losers from each process overlap, and if so, what are the implications for equity? Next is presented a detailed case study of outcome double exposure to climate change and trade liberalization in Indian agriculture. The study identifies the districts and farmers that are most likely to be negatively affected by both drier conditions and import competition. The outcomes of the two processes are shown to be systematically linked to contextual conditions that influence both exposure and the capacity to respond to each process. Global environmental change and globalization processes together reinforce uneven outcomes among farmers and rural communities, creating both "double winners" and "double losers."

Agriculture and Rural Livelihoods under Global Change

Food is a basic human need, and agriculture is regarded as a vital sector in nearly every national economy. Food security is often considered an integral part of national security, and most countries have policies and institutions to develop or protect their agricultural sector in order to ensure its viability and guarantee adequate and stable food supplies. These policies may include, for example, import duties to protect domestic producers, production subsidies, price supports, provision of infrastructure and irrigation, support for agricultural research, and insurance against crop failure due to natural hazards. In economies with little agricultural production, such as Japan, reliable access to food through trade is considered a top priority. In countries where the agricultural sector is relatively large, such as Brazil, agricultural exports represent a critical source of national income.

In addition to being a key sector for both food security and trade, agriculture is also the major source of income or livelihood for roughly half of the world's population. Despite rapid rates of rural out-migration and urban growth, nearly three billion people continue to rely either directly or indirectly on agriculture for their livelihoods. In some of the countries of southern and central Africa, as much as 65 percent of the population earn a living through the agricultural sector. In contrast, less than 2 percent of the population in the more advanced economies of Europe and North America are employed in agriculture. Yet even in advanced economies, where agricultural sectoral employment shares are relatively small, agriculture plays a powerful role in national politics and remains an integral part of many cultures and heritages (Solbrig et al. 2001). Evans et al. (2002) note that although productivist principles still dominate, agriculture in many advanced economies has begun to shift toward post-productivism, which is characterized by a shift from quantity to quality in food production; the growth of on-farm diversification and off-farm employment; dispersion of production patterns; the extensification and promotion of sustainable farming through agri-environmental policy; and environmental regulation and restructuring of government support for agriculture (Wilson 2001; Evans et al. 2002).[1]

Global environmental change and globalization touch upon nearly every aspect of agricultural production. Global environmental changes that are directly linked to agriculture include land use and land cover changes, deforestation, reductions in soil productivity and water quality, and changes in climate conditions (Tilman et al. 2001; McLauchlan 2006). The agricultural sector is also a major contributor of greenhouse gas emissions, with activities such as rice cultivation and livestock production releasing both methane and nitrous oxides. Yet at the same time, many types of agroecosystems, particularly small-scale, traditional systems, mitigate facets of global environmental change. Both small-scale traditional systems and alternative approaches—such as agroforestry in place of conventional agriculture—play a critical role in the preservation of biodiversity, maintenance of soil organic matter, carbon sequestration, and rainwater capture and retention, among other effects (Koocheki and Gliessman 2005; Pretty et al. 2006; Palma et al. 2007).

Globalization-related changes that affect agriculture include an expansion of international trade in agricultural products, reduction of domestic subsidies and supports for agricultural production, and the proliferation of new biotechnologies, including genetically modified organisms (GMOs) (Conway and Toenniessen 1999). Changes in tastes and preferences toward greater worldwide consumption of meat and dairy products, the growing influence of large supermarket chains, and the emergence of new markets for organics and other specialized products are also dramatically changing agriculture and food production systems (Atkins and Bowler 2001; Pingali 2007).

While many facets of global environmental change and globalization influence rural livelihoods, two of the most significant transformations include climate change and trade liberalization. Anthropogenic climate change not only is increasing global temperatures but is changing the amount and distribution (both spatial and temporal) of precipitation and other variables important to agriculture. In short, climate change is likely to have widespread impacts on the productivity, viability, and sustainability of agriculture in many parts of the world. At the same time, liberalization of international trade in agricultural products and the removal of domestic supports for agricultural producers are changing the economic basis for agricultural production. While trade liberalization is expected to increase production globally, it is also subjecting farmers to greater international competition, in some cases challenging the viability of the farming sector.[2] Although climate change and trade liberalization have particularly strong implications for those parts of the world that are heavily dependent on agriculture for basic livelihoods, all agricultural regions and producers will feel some effect of these two processes.

Climate Change and Uneven Outcomes

Agriculture is among the most climate-sensitive sectors in most national economies. Every aspect of production, from sowing to harvesting (and in some cases transport to markets), is influenced by climatic conditions (Parry 1990). Climate change is likely to lead to changes in the length of the growing season; loss of soil moisture resulting from changes in both precipitation and evapotranspiration; enhanced growth of some crops due to CO_2 fertilization; soil erosion from wind, heavy precipitation, and loss of snow cover; increased salinization of soils, and the introduction of new pathogens and diseases (UNEP 2006; Easterling and Aggarwal 2007). These changes will influence the types and varieties of crops that are suited to a particular region, as well as the farming practices required to grow them (e.g., irrigation demand). Climate change may also potentially lead to an increase in the frequency and intensity of extreme events and hazards, such as droughts, floods, and pest outbreaks, which is likely to have large implications for food security (Mearns et al. 1997; Sivakumar 1998; McCarthy et al. 2001). There is little doubt that climate change increases the uncertainty associated with agriculture and rural livelihoods and that it will amplify the dramatic transformations that are already taking place in this sector (Guy Robinson 2004).

The outcomes of climate change on agriculture and rural livelihoods have been assessed, evaluated, measured, and monitored in multiple ways. These include studies of the climate sensitivities of crop yields, plant competition, pollination biology, pests and pathogens, and soil conditions. Farm-level adaptation strategies and regional-scale adaptations are often incorporated in studies of the consequences of climate change on crop yields and production (Parry et al. 2004). Other types of studies focus on vulnerability and the capacity of households and communities to cope with and adapt to changing climate conditions. The approaches to understanding the outcomes of climate change can be linked to the global environmental change discourses described in chapter 2.

Studies within the biophysical discourse tend to emphasize the impacts of climate change on agricultural production. Future climate scenarios based on the results of general circulation models are often used to assess the sensitivity of agricultural production to changing climatic conditions. Process-based models simulate the impacts of climate change on crop development by integrating information about soil types, seed varieties, planting dates, and the length of the growing season with future climate scenarios (Jones and Thornton 2003; Parry et al. 2004). These models indicate that climate change will lead to both positive and negative yield changes, depending on the crop type, the scenario, and assumptions about adaptation. Some crops, particularly wheat and rice (i.e., C_3 crops that photosynthesize carbon dioxide efficiently), may benefit from climate change (Fuhrer 2003). Other crops, including C_4 crops such as maize, sorghum, and millet, are likely to experience declining yields (McCarthy et al. 2001). Modeling interactions between different types of plants and their pollinators or competitors provides an understanding of how agroecosystems will respond to climate change. There is evidence that intensive agriculture may have the potential to adapt to changing conditions, and that low-input systems may be more seriously affected (Fuhrer 2003). Combined with global economic models and international trade models, such studies provide an indication of how global production totals are likely to be influenced by climate change (Reilly et al. 1994; Fischer et al. 2002). On the whole, these modeling studies suggest that net global agricultural yields will be minimally affected by climate change (Parry et al. 2004).

However, aggregated impact studies include one very important caveat: There will be significant *regional differences* in the outcomes of climate change on agricultural production. Fischer et al. (2002), for example, found that agriculture in developed countries is likely to benefit from climate change but that developing regions, with the exception of Latin America, are likely to face negative impacts. Disaggregated analyses of the impacts of climate change on agricultural production point not only to a disparity of outcomes but to a growing divergence of outcomes: "While global production appears stable, regional differences in crop production are likely to grow stronger through time, leading to a significant polarization of effects, with substantial increases in prices and risk of hunger amongst the poorer nations, especially under scenarios of greater inequality" (Parry et al. 2004, 66). Although some studies suggest that international trade may compensate for the unequal outcomes (Reilly et al. 1994), there is a general consensus that "[p]eople who live on arid or semi-arid lands, in low-lying coastal areas, in water-limited or floodprone areas, or on small islands are particularly vulnerable to climate change" (Watson et al. 1996, 24).

The outcomes of climate change for agriculture and rural livelihoods have been approached quite differently by researchers working from other perspectives and discourses. Studies falling within both the human environment and critical discourses have explored the consequences of climate change for farmers and rural communities. This literature recognizes that the impacts of climate change will not be uniform, even within homogenous agroecoregions, because differing assets, technologies, knowledge, and other contextual factors influence farming practices and rural livelihoods (Liverman 1990; Vásquez-Léon et al. 2003: O'Brien et al. 2004; Polsky and Cash 2005). These place-based studies not only illustrate how contextual factors produce differential outcomes, but also show how contexts both affect and are affected by changes taking place at different scales (Turner et al. 2003b). Such studies emphasize that rural livelihoods will be influenced by multiple stressors, including but not limited to climate change (Luers et al. 2003; Eakin 2006; Lind and Eriksen 2006; Ziervogel et al. 2006).

The concept of social vulnerability is also used within this literature to assess how dynamic contextual factors contribute to negative outcomes (Eakin and Luers 2006). Drawing on approaches from political ecology, these studies demonstrate how evolving social and political structures, differential access and entitlements, and power relations create contexts of vulnerability within agricultural communities and households (Bohle et al. 1994; Adger 1999). Social vulnerability studies recognize that poor people are most vulnerable to climate change and that smallholders and farmers working on marginal lands will have the most difficulty adapting to climate change. Yet the consequences for agriculture and rural livelihoods are not interpreted as the inevitable result of differential biophysical changes. Rather "the risks deriving from those changes are mediated locally through social, economic, and political factors influencing the exposure of societies to such changes, and their ability to adapt" (Forsyth 2003, 176). In other words, poor people are vulnerable not simply because they are poor, but because economic, political, and social processes influence their exposure and capacity to respond to stressors and shocks (Khandlhela and May 2006).

While there is general agreement that climate change will have uneven outcomes for regions, nations, communities, social groups, households, and individuals, the identification of winners and losers, as well as the factors that are considered most responsible for such outcomes, differs significantly across the different discourses and literatures. As is discussed in the next section, there is also widespread recognition of the uneven outcomes of trade liberalization, yet disagreement about why these differences occur.

Trade Liberalization and Uneven Outcomes

Agricultural trade liberalization entails reduction or elimination of tariff and non-tariff barriers to trade in agricultural and food products, as well as reductions in national agricultural subsidies and institutional supports (Kennedy and Koo 2002).[3] As with climate change, trade liberalization is having dramatic yet uneven effects on agriculture in both developing and industrialized countries (Hertel et al. 2003; Litchfield et al. 2003; Clapp 2006; van Meijl et al. 2006). Liberalized trade not only alters patterns, strategies, and rewards associated with agricultural production, but also introduces significant volatility and variability to agricultural commodity prices, adding new uncertainties to rural livelihoods (Chand et al. 2004; Eakin et al. 2006).

The push toward liberalization of trade in agricultural products—a strategy initiated and advocated primarily by those operating within the discourse of benign globalization—is intended to increase the efficiency of agricultural production worldwide by opening national markets to international competition. The logic behind trade liberalization is the law of comparative advantage, which holds that regions should specialize in the production of those agriculture products in which they have a relative advantage, because of either climate and environmental factors or labor costs.[4] According to this logic, liberalized agricultural markets will both increase production efficiency and provide a wider variety of products at lower prices for consumers.

Advocates of free trade, particularly neoclassical economists, tend to emphasize the positive net effects of agricultural trade liberalization, including poverty reduction (Dollar and Kraay 2004; Winters 2004; Anderson and Martin 2005; Hertel and Winters 2006). Within this literature, the impacts of trade liberalization are typically measured on the basis of factors such as production, productivity, prices, price volatility, and aggregate income. Although the research acknowledges that differential or uneven outcomes emerge across agricultural regions and types of farmers, depending on cropping patterns and productivity, the differences are seen as a temporary condition. Differentials are expected to diminish over time as farmers within different regions respond to market signals and alter production and cropping patterns in accordance with international price levels. Farmers who produce crops that are not competitive in international markets are expected to shift to other, more competitive crops, or possibly even withdraw from the farming sector altogether.[5]

In contrast to pro-free trade approaches, researchers working within the discourse of malignant globalization see the uneven outcomes of trade as permanent features of an unfair system that favors large farmers, multinational agribusiness firms, and large multinational buyers, including supermarket chains. Agricultural trade liberalization is considered both as a serious threat to the livelihoods and viability of small farmers and rural communities, and as a threat to environmental sustainability (Ritchie 1993). For many small farmers around the world who barely eke out a living in agriculture, policy changes linked to trade liberalization are expected to deliver a significant blow to well-being and livelihoods. These farmers, particularly in developing countries, are not expected to be able to compete in global agricultural markets, because of structural and institutional constraints as well as systematic biases in the rules.[6] The logic of free trade suggests that these farmers should not be engaged in agriculture. However, alternative livelihood strategies may not be available without migration to other areas, including cities, where there are no guarantees for improved well-being (Halweil 2000).

Sanitary and phyto-sanitary (SPS) standards and regulations illustrate one component of trade liberalization which, although designed to protect consumers, is regarded as unfair to exporters in developing countries (Narayanan and Gulati 2002). This standardization began with the U.N Codex Alimentarius Commission (CAC), which was set up during the 1960s to create international standards for food safety in order to promote international trade (Atkins and Bowler 2001). SPS requirements, which were adopted by the WTO through its role as arbiter of sanitary and phyto-sanitary regulations, present challenges to small-scale farmers in developing countries, who cannot afford expensive equipment, cannot comply with restrictions on child labor, and cannot afford the increasing cost of transactions (Shiva 2002).[7] Retailer-dominated supply chains are similarly regarded as unfair barriers to small-scale farmers.[8] The strict requirements of supermarket chains in terms of size and quality are especially demanding in regard to information flows, capital requirements, and governance and management of the system (Kydd 2002). Producers are expected not only to meet quality criteria related to size, color, texture, and taste, but also to adjust production volumes rapidly in response to market trends, and to keep up with cost-reducing technical progress (Kydd 2002). Responding to trade liberalization has thus become increasingly complex and demanding, particularly for small-scale farmers.

Many advocates of the transformative globalization discourse argue for a more level international playing field for agricultural trade, pointing out that the United States and European Union still restrict imports of high-value and potentially more profitable agricultural products, such as sugar, meat, fruits and vegetables, and processed foods, from developing countries (IFPRI 2003). Others emphasize the need for a larger role for the state, including social welfare programs and public insurance systems, in order to compensate losers and mitigate uneven outcomes (Khor 2001; Rodrik 1997).[9] In order to reduce power differentials between multinational firms and small-scale farmers, alternative trading systems such as managed trade, fair trade, and self-reliant trade have been proposed (Dunkley 2004). Stiglitz and Charlton (2005), for example, suggest that fairer trade might also be achieved by removing barriers to imports from countries whose economies are smaller and poorer than one's own. More generally, the issue of trade liberalization has been at the forefront of efforts to promote a different type of globalization that can provide welfare guarantees for the losers from free trade while at the same time bringing the benefits of growth (Khor 2001; Stiglitz and Charlton 2005).

Some scholars and policy makers also recognize that globalization itself is changing the nature of free trade. The doctrine of comparative advantage is regarded as difficult to apply to a world that is increasingly characterized by unequal global value chains (Dunkley 2004; Friedman 2005). Changes in retail and consumption patterns are recognized as giving more weight to the symbolic and in-person service quality attributes of a traded product than to the material content. For example, Daviron and Ponte (2005) show that in the case of coffee, an emphasis on quality and service has shifted profits to the retail side of the value chain, a fact that partly explains why the coffee boom in coffee consumer societies has been accompanied by a crisis for coffee producers.

Finally, there are growing challenges to what Dunkley (2004) refers to as the "Free Market Economic Rationalist" view that links trade to development in a direct, linear, and simplistic manner. Yet many of these critics still acknowledge some benefits of trade liberalization, including the potential to improve lives and livelihoods and promote development, assuming that appropriate institutions and policies are enacted (Stiglitz and Charlton 2005, Pinstrup-Andersen 2002). As Khor (2001, 36) argues:

> What is important is the quality, timing, sequencing and scope of liberalization (especially import liberalization), and how the process is accompanied by (or preceded by) other factors such as the strengthening of local enterprises and farms, human resource and technological development, as well as the build-up of export capacity and markets.

As this brief review suggests, liberalization of agricultural trade remains a deeply contentious topic, with sharp disagreements both within and across nations as to whether and how such policies should be pursued. Although there is a general consensus that trade liberalization has uneven effects, there remains disagreement about whether the uneven outcomes are temporary and will disappear under free markets, whether they are permanent features of an unequal system, or whether they need to be fixed via assurances of fair trade.

Outcome Double Exposure and Agriculture

Climate change and trade liberalization are two global processes that are transforming the nature of agriculture. Yet, as has just been discussed, these processes will not affect every crop, location, or type of farmer equally. Climate change will lead to more dramatic and intense consequences in some places and for some farmers relative to others. Likewise, trade liberalization is a process that differentially affects those whose livelihoods are directly or indirectly related to the agricultural sector. While each of these processes has been researched and discussed extensively, there has been surprisingly little attention to how *both* processes together will affect farmers and communities. What happens, for example, when a climate-related disaster occurs at the same time that prices for livestock or marketable crops decline because of import competition? Or when subsidies for seeds or fertilizers—or agricultural extension services—disappear just when new types of crops or treatments are needed to adapt to changing climate conditions?

The double exposure framework can be used to assess how the outcomes of the two processes may overlap, and why those regions and farmers that are negatively affected by climate change also tend to be negatively affected by trade liberalization. The framework stresses that these negative (or positive) overlaps are not coincidental but instead reflect contextual conditions that influence both the nature of exposure to global change processes and the opportunities for responding to those processes. Acting synergistically, the two processes may exacerbate existing inequalities within and across rural regions. The next section illustrates the pathway of outcome double exposure through a case study of climate change and trade liberalization, focusing on uneven outcomes and implications for agricultural and rural livelihoods in India.

Climate Change, Trade Liberalization, and Rural Livelihoods in India

India is often hailed as a success story as the result of globalization (Friedman 2005). India now has the fourth largest economy in the world in terms of purchasing power parity, and it is expected to become the world's third largest economic power within the next two decades (Economy Watch 2007). At the same time, its economy is becoming increasingly polarized (Jha 2004; Milanovic 2005; Baddeley et al. 2006). Despite the growth of a highly educated, urban middle class that is connected to the global economy through jobs in high-technology industries and services, much of India's population remains dependent on agriculture, and within this group many are poor and growing poorer. Although agriculture's contribution to GDP has decreased from 30 to 20 percent over the past fifteen years, in 2001 India's agricultural sector still employed approximately 59 percent of the county's workforce (eCensus India 2002). Within rural areas, the share of agricultural workers within the labor force was 73 percent in 2001, down from 81 percent in 1991.

Income polarization in India is not simply an urban versus rural phenomenon. There are also growing differentials across agricultural areas of the country as well as among different types of farmers and laborers within those areas. Large, export-oriented farmers and regions with access to infrastructure such as advanced irrigation systems are doing relatively well, while landless laborers and regions of dryland agriculture are faring poorly. Farmers in the state of Punjab, for example, have benefited from trade liberalization while farmers in Orissa, Madya Pradesh, and Assam have been detrimentally affected (Müller and Patel 2004). As will be demonstrated in this case study, these rural differentials are likely to worsen under the combined effects of climate change and trade liberalization.

Climate Change

Climate change is a serious concern for Indian agriculture, a sector that is already highly vulnerable to present-day climate variability and change (Mall et al. 2006; O'Brien et al. 2004). A large part of India is located within the semi-arid tropics, which are characterized by low and often erratic rainfall (Ribot et al. 1996; Gulati and Kelley 1999). Much of the country relies on tropical monsoons for the majority of total annual rainfall, most of which falls within a three- or four-month period (Mitra et al. 2002). Under climate change, India could experience warmer and wetter conditions, particularly if the summer monsoon becomes more intense (Mitra et al. 2002, Kumar et al. 2002, McLean et al. 1998). While higher temperatures and growth-enhancing CO_2 effects of climate change may lead to productivity increases for some irrigated crops, the overall impact on Indian agriculture remains uncertain (Aggarwal and Mall 2002; Mall et al. 2006). Increases in monsoon precipitation and CO_2 enhancement may be offset by increased rates of evapotranspiration due to the higher temperatures, leading to negative impacts on soil moisture and crop production levels, particularly in areas that become increasingly water stressed (Dinar et al. 1998, Kumar and Parikh 2001, Lal et al. 1998, Gadgil 1995). Future water stress could also be substantially exacerbated by long-term reductions in river flows due to a reduction of the Himalayan glaciers under climate change (Kulkarni et al. 2007).

Regionally down-scaled general circulation models provide one starting point for understanding how climate change impacts will vary across agricultural regions in India (O'Brien et al. 2004). Plates 1a and 1b illustrate current and future climate sensitivity based on the results of the HadRM2 general circulation model, in combination with measures of climate sensitivity (specifically, indices of dryness and climate variability). These maps do not take into account the potential for climate-related flooding, which affects approximately 40 million hectares of land in India, causing damage to life and property (Gupta et al. 2003). Although the HadRM2 model represents only one possible scenario of climate change, it nonetheless shows that potential changes in regional climate sensitivity may occur as the result of exposure to climate change.

According to the results for the 1961–1990 period, the agricultural areas with high to very high climate sensitivity are located in the semi-arid regions of the country, including major parts of the states of Rajasthan, Gujarat, Punjab, Haryana, Madhya Pradesh, and Uttar Pradesh. Under the HadRM2 climate change scenario, all of these districts remain sensitive, but climate sensitivity for agriculture noticeably increases in Uttar Pradesh, Madhya Pradesh, and Maharashtra.

Trade Liberalization

India's agricultural sector is also undergoing a rapid transition due to trade liberalization (Bhalla 1994; Chaudhury 1998; Gulati and Kelley 1999; Sen 1999; Adhikari 2000; Eashvaraiah 2001; Gulati 2002; Vakulabharanam 2005). Agricultural trade reforms in India, which began in 1991, have included reductions and changes in import and export restrictions and tariffs, changes in access to agricultural credit, and reductions of production subsidies (UNIDO 1995, Rajan and Sen 2002). Although liberalization of agricultural trade has been limited relative to other sectors of the Indian economy, some import-competing crops, particularly oilseeds, have been fully liberalized (Sachs et al. 2000). While efforts to develop and implement the WTO Agreement on Agriculture are presently stalled, a resumption of these talks portends greater changes to come for India's agricultural sector (FAO 2003).

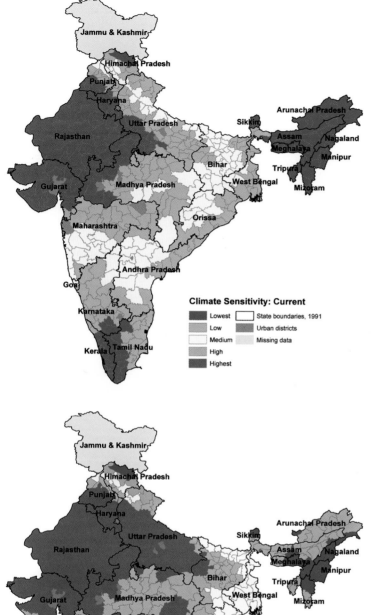

Plate 1a and Plate 1b. District-level mapping of climate sensitivity for India (1a) based on observed climate data (1961–1990) and (1b) based on results from the HadRM2 general circulation model. The districts are ranked and presented as quantiles. The same quantile breaks used in (a) were used in (b) to demonstrate absolute changes in climate sensitivity.

Climate Sensitivity: Current

■	Lowest	☐	State boundaries, 1991
■	Low	■	Urban districts
☐	Medium	☐	Missing data
■	High		
■	Highest		

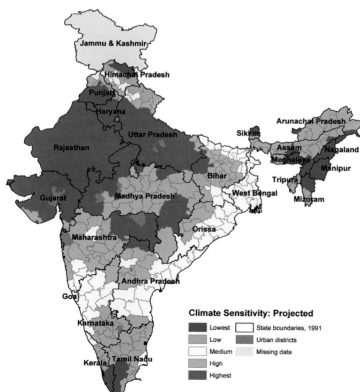

Climate Sensitivity: Projected

■	Lowest	☐	State boundaries, 1991
■	Low	■	Urban districts
☐	Medium	☐	Missing data
■	High		
■	Highest		

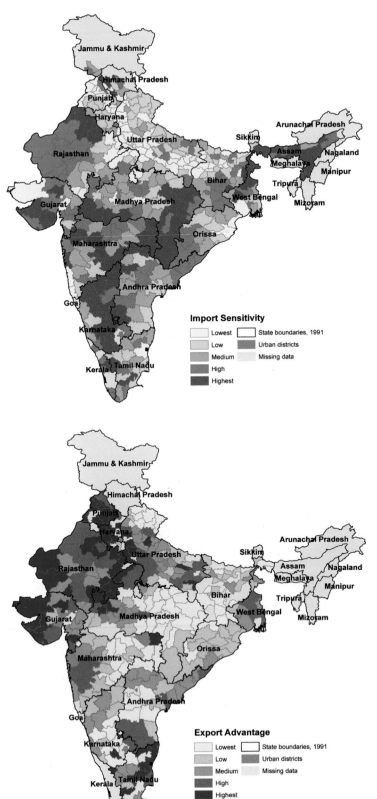

Plate 2a and Plate 2b. District-level mapping of import sensitivity (2a) and export advantage (2b), based on a representative basket of import-sensitive crops (2a) and export competitive crops (2b). Districts are ranked and presented as quantiles.

Import Sensitivity

Lowest	State boundaries, 1991
Low	Urban districts
Medium	Missing data
High	
Highest	

Export Advantage

Lowest	State boundaries, 1991
Low	Urban districts
Medium	Missing data
High	
Highest	

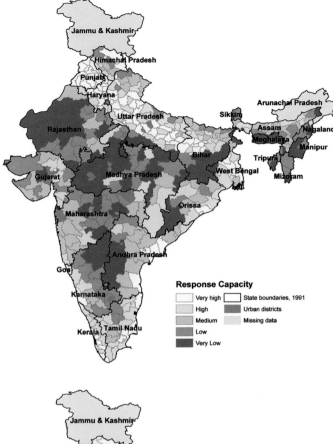

Response Capacity

Very high	State boundaries, 1991
High	Urban districts
Medium	Missing data
Low	
Very Low	

Plate 3.
District level mapping of response capacity based on a composite of biophysical, economic, social, and technological indicators. Districts are ranked and presented as quantiles.

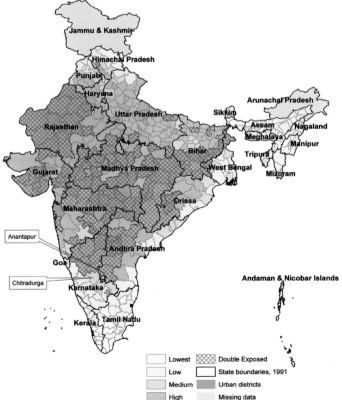

Lowest	Double Exposed
Low	State boundaries, 1991
Medium	Urban districts
High	Missing data
Highest	

Plate 4.
District level, combined map depicting both climate change sensitivity and import sensitivity overlaid on response capacity. Districts that are highly vulnerable to both climate change and import competition are considered double-exposed (identified via cross-hatching).

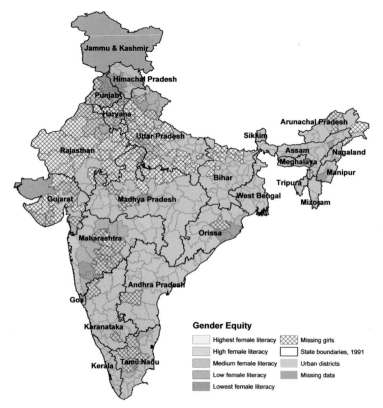

Gender Equity

☐ Highest female literacy	▨ Missing girls	
☐ High female literacy	☐ State boundaries, 1991	
☐ Medium female literacy	☐ Urban districts	
☐ Low female literacy	☐ Missing data	
☐ Lowest female literacy		

Plate 5.
District level mapping of gender equity, based on indicators of female child mortality rates and female literacy rates. Districts are ranked and presented as quantiles.

The effects of trade liberalization are likely to be highly uneven across India, with some agricultural regions and farmers benefiting from market liberalization and from new inflows of investments and technology, while others have difficulty adjusting to a more open economy, particularly to the effects of increasing competition from imports (Gulati and Kelley 1999; Mittelman 2000; Vakulabharanam 2005). Plates 2a and 2b illustrate differential regional exposure to trade liberalization based on productivity, cropping mix, and proximity to international ports—factors that are often stressed as important to trade performance within the discourse on benign globalization. As is illustrated in plate 2a, highly import-sensitive regions are found throughout the country, but they tend to be concentrated in the southern and central states of Karnataka, Maharashtra, and Madhya Pradesh and in the eastern state of Assam. All of these regions are characterized by low productivity and a higher than average concentration of production in import-competing crops, particularly oilseeds. Regions that are likely to be able to take advantage of export opportunities are illustrated in plate 2b. These regions of "export advantage" tend to have higher productivity and better proximately to coastal port areas; they include Punjab, West Bengal, Andra Pradesh, and Tamil Nadu.

Uneven Outcomes across Agricultural Regions

Exposure to both climate change and trade liberalization will vary across different regions in India. Nevertheless, the processes alone do not determine outcomes. The pathway of outcome double exposure recognizes that contextual conditions differentially influence both exposure and response capacities. Thus some agricultural regions may be highly vulnerable to small changes while others are able to cope with or respond positively to large changes. As is emphasized in some of the global change discourses, the sources of these variations in response capacity are rooted in factors such as access to technology, social capital, and institutional capacity.

Differences in contextual conditions across Indian regions are illustrated in plate 3, which maps a composite of socioeconomic, biophysical, technological, and institutional factors that are likely to influence capacity to respond to both climate change and trade liberalization. This map depicts higher degrees of response capacity (i.e., more favorable contextual conditions) in districts located along the Indo-Gangetic Plains (except Bihar) and lower response capacity in the interior portions of the country, particularly in the states of Rajasthan, Madhya Pradesh, Maharashtra, Andhra Pradesh, and Karnataka (O'Brien et al. 2004).

Within the double exposure framework, both differential exposure to global change processes and differential response capacities influence outcomes. To illustrate the overlapping outcomes of each process, information from the response capacity map was first combined with the climate change and import sensitivity maps to create two vulnerability maps (depicting vulnerability to climate change and vulnerability to import competition). Superimposing the two maps (plate 4) reveals those areas that are highly exposed to each process *and* have weak response capacity due to unfavorable contextual conditions. The map identifies districts that are vulnerable to negative outcomes from both climate change and import competition.

These double exposed districts (highlighted via crosshatching in plate 4) are concentrated in Rajasthan, Gujarat, and Madhya Pradesh as well as in southern Bihar and western Maharashtra. Within these districts, trade liberalization and climate change are likely to pose simultaneous challenges to the agricultural sector.

Reacting to two processes of change will, of course, present challenges throughout India, but farmers in the double exposed districts are likely to feel disproportionately more stress, particularly if there is a mismatch between climate-compatible crops and market-driven demand for those crops. It is in these areas of double exposure where policy changes and other interventions may be most needed in order to help farmers negotiate changing contexts for agricultural production.

Uneven Outcomes for Communities and Farmers

While cross-regional examination of outcome double exposure helps to identify spatial areas that are likely to be negatively affected by both climate change and trade liberalization, others lenses of analysis are also valuable. Here we consider outcome double exposure for communities and farmers located in two neighboring districts (and states): Anantupur (Andhra Pradesh) and Chitradurga (Karnataka) (O'Brien et al. 2004).[10] This approach highlights how differing contextual conditions in the two districts influence responses to global change processes. In particular, it illustrates the importance of infrastructure and irrigation in responding to agricultural change, and it shows the differential effects of climate change and trade liberalization on the horticultural and oilseeds sectors (Chand 2004).

According to regional analysis presented above, the two districts have similar levels of exposure to both trade liberalization and climate change. The relatively high exposure to import competition in both cases is largely a function of the concentration in production of groundnuts. The reduction of oilseed import restrictions in India, which began in the middle 1990s, introduced significant price competition for Indian oilseed producers because of imports of inexpensive Malaysian palm oil (Shiva 2000). Similar levels of exposure to climate change (low to medium) in the two districts stem from similar current climatic conditions (e.g., temperature and rainfall patterns) and projected model effects. Both districts are also subject to comparable levels of climatic variability, having both recently suffered from multi-year droughts, which ended in 2003.[11]

Yet the two districts exhibit some differences in contextual conditions. Anantapur can be considered to have a weaker response capacity than Chitradurga for several reasons. In terms of technology and infrastructure, Chitradurga has greater access to irrigation than Anantapur, which is primarily a rain-fed agricultural area. Institutional differences include a higher presence of state institutions and NGOs in Chitradurga than in Anantapur. Table 5.1 compares several measures of the social context, including how they changed in the decade following liberalization. Anantapur has lower levels of both general and female literacy than Chitradurga as well as a higher percentage of landless laborers. These contextual conditions appear to be critical to the capacity to respond to global change processes.

Table 5.1
Social conditions in the case study districts (percentages)

District	Chitradurga		Anantapur	
Year	1991	2001	1991	2001
Workforce employed in agriculture	83.91	81.75	84.96	79.96
Literacy of the population	48.69	53.73	38.28	45.33
Literacy of the female population	35.42	44.17	21.10	33.35
Landless laborers within the agricultural workforce	46.03	47.84	51.02	54.98

Sources: Census of India, 1991; Census of India, 2001

As a result of these differing contextual conditions, farmer responses to import competition and climate variability (i.e., drought) in the two districts are very different. Free-market economics would suggest that farmers in both districts should respond to import competition and drought by shifting to production of economically viable and drought-tolerant crops. Yet for dryland farmers in Anantapur, efforts to shift to production of other types of drought-tolerant crops are constrained by institutional and marketing barriers. In particular, other rain-fed crops, such as different fruit varieties, either require too much capital or do not have a long enough shelf life to be marketable, given the current transportation and storage infrastructure. Without irrigation, water-harvesting systems, or alternatives to groundnuts, dryland farmers in Anantapur have few options for responding to either climate change or trade liberalization (O'Brien et al. 2004).

By contrast, farmers in Chitradurga have a wider range of response options. Because irrigation is available, many farmers in this district have been encouraged through state government and private initiatives to cultivate alternative crops, such as areca-nut, pomegranate, and banana. Over recent years, export companies have increasingly entered into buy-back contracts with farmers for gherkin production aimed at European markets, with plans to expand to other vegetables. Although a wider range of adaptation strategies were available to farmers in Chitradurga than to those in Anantapur, it is the larger farmers who tend to benefit from government subsidies for drip irrigation, sericulture rearing houses and other production technologies, formal bank credit, crop insurance, and access to larger markets. Smaller farmers are typically disadvantaged because of lack of information and their dependence on local merchants for credit. Furthermore, irrigation, which contributes to higher response capacity within the district, may not be sustainable in the long run, particularly if water-intensive horticultural crops are produced for international markets while water availability is reduced because of climate change (O'Brien et al. 2004).

This case study makes it clear that adaptation to both globalization and climate change at the local level in India requires more than simply behavioral changes by producers (i.e., it is not just a matter of enhanced social capital). This finding, which is confirmed by other research—such as the work of Deshingkar et al. (2003) on horticulture marketing in Andra Pradesh, and of Storm (2003) on the distributional consequences of liberalization across India—suggests that institutional supports, both governmental and nongovernmental, for infrastructure and marketing are vital to ensure that both large and small farmers benefit from liberalization under climatic change. The results further suggest that state-level agricultural policies, which vary across India, may play a critical role in increasing local adaptability to climate variability and change in the context of trade liberalization. In the case of Anantapur, institutional barriers leave farmers who are double exposed poorly equipped to adapt to either of the stressors, let alone both of them simultaneously. In Chitradurga, on the other hand, institutional support appears to facilitate adaptation to both climatic change and trade liberalization. However, these supports tend to disproportionately benefit the district's larger farmers (O'Brien et al. 2004).

Uneven Outcomes among Rural Women

The double exposure framework may also be used to assess uneven outcomes across different types of individuals and social groups. In India, an overwhelming proportion of the rural poor belong to lower castes, scheduled tribes, and female-headed households (Müller and Patel 2004).[12] Not surprisingly, these groups are more likely to be adversely affected by environmental degradation of any form, relative to others (Roy and Venema 2002). Rural women in particular have been identified as being potentially more vulnerable to both climate change and trade liberalization, relative to men.[13] There are numerous reasons for this vulnerability, many of which have been discussed in the literatures on climate change, globalization, and rural development (Ellis 2000; Mittelman and Tambe 2000; Pande 2000; Nelson et al. 2002; Denton 2002; Piana and Lambrou 2005). These include prevailing inequalities in access to land, control over resources, ability to command and access paid labor, and decision-making possibilities. At the same time, there has been a "feminization of agriculture" in part because men tend to migrate to urban areas in search of work, leaving women to carry out agricultural work on their own (Bruinsma 2003). In many cases women's vulnerability is closely linked to poverty, including lower incomes and fewer opportunities for off-farm employment, as well as to their dual roles in agricultural production and the social reproduction of households (Skutch 2002; Vincent 2007).

In India, rural women in general have few rights to land, and traditional usufruct rights to community land were lost after land reform (Roy and Venema 2002). Land, according to Agarwal (1998), not only provides a source of economic security (through direct production possibilities and as a mortgageable or sellable asset during crisis), but also defines social status and political power in the village, structuring relationships both within and outside of households:

> Yet for most women, effective rights in land remain elusive, even as their marital and kin support erodes and female-headed households multiply.

In *legal* terms, women have struggled for and won fairly extensive rights to inherit and control land in much of South Asia; but in practice most stand disinherited. Few own land; even fewer can exercise effective control over it.[14] (Agarwal, 1998, 2).

Furthermore, a decline in farm-level income has large effects on rural women, who are among the lowest-paid agricultural laborers. The number of women sustaining their livelihoods as laborers rather than cultivators has grown as the result of mechanization and technical interventions, which they are traditionally excluded from using (Roy and Venema 2002). Indeed, Müller and Patel (2004) found evidence that women are increasingly employed in labor-intensive agricultural production in export-oriented areas at lower wages than men, and also taking over labor-intensive activities on their own farms. Yet they are the first to be fired with mechanization of agriculture, and they have faced reduced opportunities for off-season employment. From their analysis, Müller and Patel (2004, 52) conclude that "women have become the disposable factor of production in Indian agriculture." For these reasons, it can be expected that climate and trade-related stresses and shocks many have disproportionately negative outcomes on women in India.

In an analysis of outcome double exposure in India, gender equity may be considered an important contextual factor that influences both exposure and response capacity. Measures of gender equity provide an indirect indicator of women's access to land, resources, and decision-making capabilities. In plate 5, gender equity is calculated on the basis of indicators of female child mortality rates and female literacy rates (see Drèze and Sen 2002; Sen 1999). Districts with female child mortality rates in excess of natural rates and districts with lower female literacy rates have lower values for gender equity. Women living in districts with higher degrees of gender equity may be better able to respond to shocks and stressors and to take advantage of new opportunities. It must be recognized, however, that gender equity in select social indicators does not always amount to equity in gender relations, and in particular it does not necessarily equate to equal access to land and property rights (Kodoth and Eapen 2005).

While both rural men and rural women may be confronted by the same processes in India, contextual differences can lead to differing degrees of exposure, responses, and outcomes. However, gender inequity may also be a *result of* overlapping outcomes from these two processes. That is, the differential effects of global change processes may be both observed and perpetuated across gender groups. While the gender equity map suggests that rural women in many districts in India will likely be made worse off as the result of both processes, further analysis of outcome double exposure at the household and individual levels within specific districts would shed light on the key factors that enable or constrain women's responses to both processes.

Emerging Contradictions

The outcomes of global change may be observed through many different lenses in India (see figure 5.1). In each case, the pathway of outcome double exposure

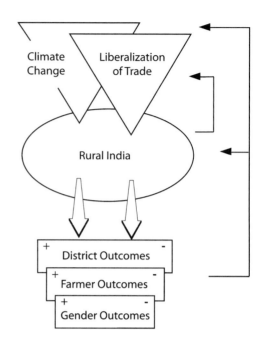

Figure 5.1
Outcome double exposure in Indian agriculture. Uneven outcomes may be observed by district, by type of farmer, and by gender.

demonstrates that the uneven outcomes of one process cannot be addressed without one's taking into account the other process. In many districts, climatic changes may limit the ability of farmers to adapt to new market conditions in a globalizing agricultural sector, just as agricultural trade liberalization may reduce the capacity to respond to climate change. Schemes to expand irrigation to help small farmers participate in export markets, for example, are likely to be doomed if they do not account for the potential for changes in water availability as the result of climate change. Institutional changes such as removal of quotas or reductions in tariffs under trade liberalization may open new opportunities for export sales yet may make it more difficult for many farmers to adapt to climate variability and change if markets for traditional drought-tolerant varieties are destroyed by import competition.

In addition to multiple, overlapping processes, another insight from the India case study is that there is a need to take into account multiple pathways of double exposure. Although this chapter has emphasized outcome double exposure, context and feedback double exposure may also be seen within Indian agriculture. Increased import dependency for staple crops such as wheat and rice, for example, may change the context for agricultural production, introducing new uncertainties and vulnerabilities for Indian food security. Under liberalized trade, the prices and availability of staple crops may increasingly depend on the sensitivity of other food-producing nations to climate change. At the same time, the expansion of irrigation water usage within India in response to both climate change and trade liberalization is likely to create feedbacks that contribute to processes of global environmental change, including groundwater depletion due to accelerated rates of withdrawal and land degradation as the result of water logging and salinization. This degradation may affect the capacity to cope with and respond to all type types of shocks and stressors, thus increasing future vulnerability.

Conclusion

Outcome double exposure draws attention to a deepening relationship between global change and inequality. In relation to agriculture and rural livelihoods, sharp contrasts in human well-being and environmental conditions are already evident in rural regions throughout the world (UN 2005a; UNDP 2005; World Bank 2006b). These inequalities not only contribute to food insecurity and rural out-migration, but also are associated with political conflicts and social unrest. Furthermore, rural inequalities are increasing as a result of both globalization and global environmental change. By reinforcing inequalities and polarization, double exposure not only may jeopardize individual lives and community well-being in rural regions, but may ultimately undermine efforts to achieve social justice and promote human development.

The creation of double losers from both climate change and trade liberalization reveals one of the many challenges that these two processes raise for long-term human security. Although the overlaps and linkages between the two processes may result in some rural regions and farmers emerging significantly better off (at least in the short run), others are becoming increasingly less able to cope with or adapt to both global change processes. Addressing these challenges begins with recognizing how multiple processes of global change affect agricultural regions, and with identifying what types of policy changes can enhance the response capacity of farmers and rural communities.

Changing Contexts and Emerging Vulnerabilities

6

A new kind of capitalism, a new kind of economy, a new kind of
global order, a new kind of society and a new kind of personal life
are coming into being, all of which differ from earlier phases of social
development.
—Ulrich Beck, *World Risk Society*

Change is an important and inevitable part of life. Yet the nature of change has
transformed dramatically in recent decades. Writing in 1970, sociologist Alvin
Toffler drew attention to the notion of "future shock"—the shattering stress and
disorientation that we induce in individuals by subjecting them to too much change
in too short a time (Toffler 1970). In the years since Toffler coined the term "future
shock," global environmental change and globalization have emerged as two of the
core processes that are exposing people to new and unexpected shocks and stresses.
Indeed, increasing risk and uncertainty are considered by some to be hallmarks of
both types of global change (Beck 1992). Moreover, the transformational nature of
these two global processes is not simply about exposure to faster, more frequent, or
more widespread changes. It is also about the ways that the processes are influencing the vulnerability of individuals, communities, and regions to future shocks of
many types.

This chapter draws on the pathway of context double exposure to explore
how global environmental change and globalization are affecting exposure and the
capacity to respond to extreme events in urban areas. Cities have long been recognized as sites where environmental hazards, economic crises, or other acute events
often lead to disasters (Mitchell 1999b; Pelling 2003a). As locations of concentrated
development and population, cities are highly vulnerable to disruptions of various
types, including natural events such as earthquakes and hurricanes, technological

shocks such as chemical spills and power outages, and economic shocks such as currency devaluations (Vale and Campanella 2005).[1] These shocks have direct effects on the residents of urban areas through flooding, mudslides, contamination, extended periods of heat stress, or periods of economic inactivity and unemployment. They also have indirect and integrative effects through damage to water supply systems, damage to power grids, disruption of transport, and disruption of urban economic and social functions (J. Kenneth Mitchell 2004; Solecki and Rosenzweig 2007). While both global environmental change and globalization may increase the likelihood of extreme events in cities, these processes are also changing the conditions or characteristics that make particular places or groups vulnerable to shocks of different types.

Although we emphasize here how the urban context is changing as a result of double exposure, it is important to recognize that urbanization itself is a reflection of both types of global change (Sanchez-Rodriguez et al. 2005). Urbanization has become the dominant feature of human settlement patterns over the past century. More than half of the world's current population lives in cities. By the year 2015 there are expected to be 60 megacities in the world, each with a population of 10 million or more people. By 2030 more than 60 percent of the world's population is expected to live in urban areas. The largest urban population changes are expected to occur in coastal areas, particularly in Asia and Africa (Sanchez-Rodriguez et al. 2005; UN 2006).

Urban development and urban spatial expansion also represent critical drivers of global environmental change (Cronon 1991). The rapid rate and scale of urbanization contribute to global environmental change through changes in land use, the creation of urban heat islands, air pollution, greenhouse gas emissions, and intensive resource use. Globally, urban expansion consumes more than half a million acres of arable land per year (HABITAT 2001). Other changes associated with urban expansion, such as loss of wetlands and degradation of watersheds, are also occurring on a worldwide scale and similarly contributing to changes in the global environment. The environmental consequences of urbanization are not restricted to cities but often extend to other areas (Gleeson and Low 2000). These can be measured by the concept of ecological footprints—which refers to the land area and aquatic resources required to sustain an urban population (see Wackernagel and Rees 1996). For example, the ecological footprint of Oslo, Norway, is estimated to cover an area that is 90 times larger than the area of the city itself (Aall and Norland 2002).

Urbanization also contributes to many facets of globalization. Cities represent critical nodes in the global economy, facilitating concentration of production capital, growth of port and trade facilities, and the transmission of globalized consumption desires and aspirations to new residents (Solecki and Leichenko 2006). Major cities such as New York, Los Angeles, Tokyo, London, and Paris are often characterized as "central hubs" or "control points" in global production and consumption networks (Sassen 1991; Abrahamson 2004). These global cities are distinguished by disproportionate shares of multinational corporate headquarters, international financial markets and institutions, and advanced producer services, all of which contribute to the concentration of economic command-and-control functions in these cities (Sassen 1991; Yeoh 1999). On the consumption side, cultural industries such as publishing, advertising, and film are also concentrated in large cities, enhancing their role as producers and transmitters of an ideology of consumption (Sklair 2002; Princen et al. 2002; Abrahamson 2004).

This chapter points to the ways that transformations of contextual conditions in cities are increasing the vulnerability of some urban residents while also decreasing the resilience of cities to future shocks. First we will consider the urban context and discuss how factors such as highly concentrated populations, the presence of critical economic functions, and reliance on external resources create vulnerability to extreme events that often leads to disasters. Then we will look more specifically at how global environmental change and globalization are affecting the contextual environment in cities. In particular, we will see how these processes are altering biophysical, institutional, and economic conditions in cities. The pathway of context double exposure will be illustrated through a case study of Hurricane Katrina in the United States. In New Orleans, as in many other coastal urban areas in the world, global change processes have been transforming conditions, creating the context for disaster. We will see that New Orleans has been systematically becoming less resilient to shocks, as was revealed by Hurricane Katrina. Confronting future shocks requires attention to the changing nature of these events as well as a better understanding of the changing contexts within which they will occur.

Disasters in an Urban Context

From flooding in Mumbai and earthquakes in Istanbul to financial crises in Bangkok and Buenos Aires, shocks and extreme events in urban areas are increasing not only in frequency and magnitude but also in their impacts and costs. Drawing attention to the co-evolution of urbanization and urban natural disasters, Mitchell (1999a; 1999b) notes that as cities grow and expand, the scope and magnitude of urban disasters have been increasing. Others suggest that urbanization processes are reflexive: they create their own risks by causing degradation of the local, regional, and global environments (Hardoy et al. 2001; Pelling 2003a; 2003b; Sanchez-Rodriguez et al. 2005). Concentration of resources and people also means that the economic, social, and environmental costs of extreme events are high in cities (Klinenberg 2002; Kraas 2003; Pelling 2003a; J. Kenneth Mitchell 2004). These costs are likely to escalate as a result of growing populations in coastal cities, many of which are already highly vulnerable to sea level rise, tsunamis, and other hazards (Klein and Nicholls 1999; Sanchez-Rodriguez et al. 2005; McGranahan et al. 2007; O'Brien and Leichenko 2007).

There is a rich set of literature on urban hazards that can be used to explain how and why cities are becoming more vulnerable to natural disasters. This literature, which shares a common intellectual heritage with the global environmental change discourses reviewed in chapter 2, explains hazard vulnerability via different causal factors, including biophysical, geological, and technological conditions, institutional failures, or social and political inequalities (Cannon 1994; Hewitt 1997; Mitchell 1999b; Mileti 1999; Kraas 2003; Pelling 2003a; Hilhorst 2004; O'Neill 2006).[2] While there is a consensus across this literature that hazards events themselves do not automatically create disasters, each of these subsets points to different factors that make cities "crucibles of hazards" (Mitchell 1999a).

Biophysical, geological, and technological explanations, for example, emphasize the role of high-risk areas (e.g., fault zones), high concentrations of population and physical capital in small areas, a dependency on large-scale infrastructure projects, an increased risk of disease transmission due to crowded conditions, and location in fragile or vulnerable environments, particularly coastal areas (Mitchell 1999b). These explanations are often associated with a hazard-centered paradigm that focuses on the geophysical processes underlying disaster and that seeks to control outcomes through monitoring and predicting as well as through engineering projects and technologies that contain their effects (Hilhorst 2004). Post-disaster recovery policies often seek to reduce future exposure rather than to address contextual factors that influence vulnerability (Ingram et al. 2006).

Institutional explanations, in contrast, emphasize political and bureaucratic governance practices, management approaches, and response plans (Hilhorst 2004, O'Neill 2006). Other literature suggests that urban disasters are the result of the interplay of many contextual factors, including both biophysical factors and social and political inequalities which generate unequal exposure to risk, making some people and places more prone to disaster than others (Cannon 1994; Bankoff et al. 2004; Roberts and Parks 2006). From this perspective, natural disasters can be considered "signifiers of the inequalities that underpin capitalist (and alternative) development, of unsound and manifestly unsustainable human-environment relations" (Pelling 2003a, 6–7). The increase in frequency and magnitude of urban natural disasters is seen as a function both of humans pushing against the limits and constraints of the natural environment and of institutions and socioeconomic and political structures that produce differential vulnerability and differential ability to adapt (Pielke and Sarewitz 2005).

Recognition of hazard threats in cities has also engendered a growing literature on urban resilience (Godschalk 2003; Vale and Campanella 2005; CSIRO et al. 2007). For urban social-ecological systems, resilience entails the ability to recover from all types of perturbations with minimal loss or damage and without requiring a large amount of outside assistance (Godschalk 2003; Mileti and Galius 2005). While much of the urban resilience literature emphasizes hazard preparedness and recovery, urban resilience is also thought to entail flexibility of urban social-ecological systems in the face of various types of uncertainty as well as an ability to capitalize on unexpected opportunities (CSIRO et. al. 2007). Steps to ensure urban resilience to hazards may encompass a wide range of contextual changes, including, for example, land use planning to minimize risks, enhancement of social support networks, improved emergency warning systems and institutional response capacity, better infrastructure and enforcement of building codes, and broader provision of insurance (Godschalk 2003; Mileti and Galius 2005).

Parallel to the large literature on hazards in cities is a smaller set of literature on economic shocks in cities. Recognizing the role of globalization—particularly internationalization of financial markets—in producing economic shocks, this literature provides insights into the vulnerability of cities and urban residents to unexpected economic events such as currency devaluations or disinvestment due to loss of a major employer. Like the hazards literature, the economic shocks literature recognizes that shocks do not affect all cities equally, nor do they affect all urban residents equally (Rakodi and Lloyd-Jones 2002). Cities with economies that are highly concentrated in a single economic sector or industry, such as steel or automobiles, are more likely to experience significant downturns following an unexpected event such as a spike in oil prices or the closure of a large production facility.

Among urban residents, lower-skilled workers and, in general, the poor are most likely to be subject to the negative effects of economic shocks, because they have more limited livelihood options and fewer assets on which to rely (Moser 1998, Skoufias 2003, Miller 2005). Explanations for differential vulnerability to economic shocks for urban residents emphasize both individual assets such as wealth and education, as well as contextual factors such as institutions and diversity of livelihood options (DFID 1999; Rakodi and Lloyd-Jones 2002).[3] Kirby (2006) notes that it is not the increase in risk associated with globalization that creates vulnerability, but rather the decline in coping mechanisms for facing and surviving such risks.

The hazards and economic shocks literatures each acknowledge that processes of global environmental change and globalization are making extreme events more likely. Disasters, particularly in urban areas, are seen as increasing in both impact and scope because of the combined effects of environmental, economic, social, demographic, and technological changes (Kraas 2003; Oliver-Smith, 2006). Urban shocks and extreme events are also seen as more frequent and more intense as the result of climate change and the instability of global financial markets (Adger and Brooks 2003; Dore and Etkin 2003; Kraas 2003). Climate change, in particular, is seen as changing the nature of some hazards, particularly hydrometeorological disasters such as floods, windstorms, and droughts (Rosenzweig and Solecki 2001; Adger and Brooks 2003; Dore and Etkin 2003). Warmer temperatures as a result of climate change are also expected to intensify the hydrological cycle, resulting in higher variability and greater extremes (IPCC 2007b).

Similarly, globalization is making economic shocks more likely because of increasing flows of capital and speculative investments, currency and exchange rate fluctuations, and turbulence in global supply chains (Kirby 2006; Dicken 2007).[4] Furthermore, because markets centered in New York, London, Tokyo, Mexico City, Jakarta, Johannesburg, and other cities are tightly linked as part of a complex global financial system, economic shocks in one location have the ability to manifest quickly in other locations. These connections are fortified by large populations of immigrants, travel and tourism, and educational exchanges, so that a shock in one area will often have ripple effects in many other locations (O'Brien and Leichenko 2007).

Although a number of studies draw attention to the potential consequences of global change-related events for cities, less attention has been given to the ways that both of these processes are fundamentally transforming the context in which shocks and stresses are experienced. As is shown in the next section, the two processes interact through contextual changes to influence the vulnerability of individuals, communities, and social groups, ultimately affecting the resilience of urban areas.

Transforming the Urban Context: Global Environmental Change and Globalization

From the preceding discussion of the urban hazards and economic shocks literatures, we can identify some of the critical contextual factors that influence urban vulnerability to extreme events, including biophysical, social, economic, and institutional conditions. Recall that changes to many of these factors were identified in chapter 3 as key outcomes of both global change processes (see tables 3.1 and 3.2). Nearly all of the contextual factors that the hazards and shocks literatures identify as important influences on exposure and responses to extreme events are presently in flux as the result of both global environmental change and globalization This section explores three of the most prominent types of contextual transformations in cities: biophysical changes, economic changes, and institutional changes.

Biophysical Changes

The biophysical context of urban areas includes elements of the physical environment such as land, air, water, and various flora and fauna, as well as the ecological and physical systems that link these elements. These elements and systems provide critical life-supporting ecosystem services, such as ensuring adequate freshwater supplies, space cooling, and water purification. Many of these systems help to buffer urban areas from extreme events such as heavy rainfall, heat waves, and droughts. These systems also provide natural protection from hazards such as coastal storms, wildfires, and mudslides.

Urban biophysical environments are presently undergoing dramatic changes in conjunction with both globalization and global environmental change. The most visible of these changes are related to the physical expansion of urban areas. Urban expansion typically entails conversion of agricultural lands and natural habitat areas to residential and industrial uses. In addition to reduction of green spaces around cities, urban expansion often causes deforestation of upland areas, draining and filling of wetlands, and degradation of regional watersheds due to the expansion of impervious surfaces such as roads, parking lots, and rooftops (Marcotullio 2001; Cieslewicz 2002). In coastal regions, urban expansion may also entail development and modernization of port facilities. These efforts, which facilitate international trade, typically entail dredging of port areas and large-scale construction projects in ecologically sensitive wetlands and delta zones.

These changes contribute to an overall reduction in ecosystem services in and around urban areas (Millennium Ecosystem Assessment 2005), making cities less resilient to naturally occurring extreme events such as heat waves, floods, and droughts. The loss of green areas and trees reduces space cooling, including the capacity to mitigate urban heat island effects and heat waves. Damage to and loss of wetlands in coastal cities compromise functions such as flood control and purification of water supplies and reduce buffering from storms. Land degradation affects soil retention and absorption, which help to regulate water availability and ensure adequate supplies during times of drought. Deforestation on hillsides and slopes surrounding cities, particularly to accommodate new settlements, contributes to erosion, runoff, and new drainage patterns, which can result in catastrophic mudslides triggered by heavy precipitation. This was the case when tens of thousands died in the 1999 mudslide in Caracas, Venezuela (Leichenko and Solecki 2006).

Along with land use changes, another major force that is transforming the biophysical context of urban areas is climate change. Higher temperatures and altered precipitation patterns may lead to changes in water availability, which is already a significant concern for cities throughout the world (Hardoy et al. 2001). Higher average temperatures may also exacerbate urban heat island effects or extreme heat events in cities (Solecki and Rosenzweig 2007). In addition to changes in precipitation and temperatures, sea level rise is also a significant threat, particularly for low-lying coastal cities. Over the next century, sea levels are expected to increase significantly from 1990 levels (IPCC 2007b; Rahmstorf 2007). Some of the increase can be attributed to the thermal expansion of oceans, and some is linked to the melting of glaciers and ice on land (Church et al. 2001; Meehl et al. 2005; Alley et al. 2005).[5] Sea level rise may also increase the impacts of storm surges associated with extreme weather events. Storm surges are responsible for a significant amount of damage in coastal cities. Wetlands, coral reefs, and coastal ecosystems often buffer coastal cities from such extremes, absorbing much of the energy and impact.

These dramatic environmental transformations to urban areas are often emphasized in the biophysical discourse on global environmental change, as well as in the geophysical or hazard-based approaches to understanding disasters. As has been mentioned, these approaches place a heavy emphasis on monitoring, modeling, predicting, and managing environmental changes, often through technological interventions, information and early-warning systems, and evacuation plans. What is relevant to all discourses, however, is that not only are hazards changing, but the biophysical context in which they are occurring is changing as well. This change affects exposure, which in and of itself can influence the capacity to respond to shocks.

Economic Changes

The economic context in urban areas is also undergoing dramatic change. This changing context may be summarized via two contrasting realities associated with cities. One entails the dramatic accumulation of wealth in cities, as manifested through construction of gleaming office towers, gentrification of inner-city areas, and the spread of private gated residential communities. The employees and residents of these communities typically have preferential access to urban services, infrastructure, and amenities and are relatively shielded from environmental hazards. The other reality of urban areas includes the growing concentration of new rural and international migrants, the expansion of the informal sector, the lack of adequate and affordable housing, and the growing portion of residents who live in highly polluted and ecologically fragile environments.

Growing polarization and rising inequality in cities are linked with several facets of globalization, including increasing mobility of production capital, rising levels of foreign direct investment, and liberalization of trade. In cities in East and South Asia (e.g., Shanghai, Kuala Lumpur, Singapore, and Mumbai) new foreign investments and rising exports have fostered job growth and increasing wages in many sectors, particularly manufacturing (Olds 2001). In older industrial cities of North America and Western Europe (e.g., Buffalo [New York] and Manchester, UK) and in many cities in developing countries, capital mobility under globalization has the opposite effect, with industrial restructuring and the closure or relocation of production facilities to other areas leading to a large number of job losses and population out-migration (Meikle 2002; Perrons 2004). Although some of these cities have found a niche in the changing global economy, many others are struggling to adjust to new economic conditions (Beall 2002).

It is not only the changing location of manufacturing jobs that has changed economic conditions in urban areas. Globalization-related technological innovations have also contributed to a relocation of service sector jobs, particularly those that entail routine information and data processing (e.g., insurance claims, software production) to large cities such as Bangalore, while at the same time redefining them in North American cities such as New York (Friedman 2005; Blinder 2006). While these shifts bring higher wages to the most skilled workers within these sectors, they have a negative effect on displaced workers in North American cities, as well those residents of Asian cities who lack the requisite skills for the "new economy." Such workers rely instead on either low-paying service-sector jobs or jobs in the informal sector, working casually or as daily laborers with no job or income security. These changes reflect the latest twist in what has been termed a "hollowing out" of advanced urban economies, that is, a loss of middle management and blue-collar manufacturing jobs accompanied by growth in jobs at both the top and bottom of the wage spectrum. These changes in cities illustrate the concepts of flexible labor markets and flexible production, whereby low-wage workers are seen as the flexible labor force for mobile firms (Peck 1996).

Finally, growing inequality in many urban areas is rooted in demographic shifts associated with globalization. The promise of new and higher-paying jobs is drawing growing numbers of internal and international migrants to cities. At the same time, globalization-related economic changes in rural areas, including liberalization of agricultural trade (see chapter 5), have transformed livelihood opportunities, resulting in increased rural out-migration. These new migrants may come to cities seeking employment in new production facilities and the service sectors, or they may come with the intention of finding any type of employment that provides an income. The new migrants are, in and of themselves, sources for urban economic growth, economic vitality, and the transmission of new cultural ideas (Smith 2001; Katharyne Mitchell 2004). However, many migrants from rural areas end up in the informal labor market in temporary or short-term jobs, with few benefits and no income security (UNDP 2005). In some cases, particularly in less-developed regions, these jobs involve marginal, often semi-legal or illegal activities such as begging, waste picking, or prostitution (Meikle 2002). When extreme events such as floods occur, many workers in the informal sector become unemployed for months on end and often incur significant debt and other hardships (Rashid 2000).

The changing economic makeup of many cities under globalization translates into greater income disparities between relatively high-paying manufacturing and high-skill service sector jobs, and low-paying, low-skilled jobs in either the service or informal sectors. Although explanations for these changes vary among the globalization discourses, from short-term market failures within the benign discourse to systematic inequities in the capitalist system within the malignant discourse, there is general agreement that cities are experiencing an exacerbation of socio-economic inequalities and a sharpening of spatial disparities in economic and social well-being. These growing inequalities in turn are influencing patterns of exposure to hazards and extreme events. Increasing housing costs and growing competition for land in many cities means that more and more poor urban residents are left with few options except to live in ecologically vulnerable areas such as hillslopes (Davis 2006). At the same time, reduced wages and limited livelihood options mean that many people have difficulty responding to shocks. Shrinking government supports and reduced social services for low-income populations exacerbate negative outcomes. As will be shown next, the loss of these safety nets can be attributed to institutional changes associated with globalization.

Institutional Changes

Institutions represent "sets of rules, decision-making procedures, and programs that give rise to social practices, assign roles to the participants in these practices, and guide interactions among the occupants of individual roles" (Young 2002, 5). They also reflect societal norms and expectations as to what a state's responsibilities are to its citizens. In many countries responsibility for protection and recovery from extreme events has typically been regarded as falling under the purview of the state.[6] Yet the capacity to fulfill this function has been weakened by both economic and institutional changes associated with globalization (Eakin and Lemos 2006). Consequently, "In some countries, disasters are increasingly seen as the implicit breach of a social contract where states should protect their citizens from vulnerability to disaster" (Hilhorst 2004, 61).

For urban areas, institutional changes associated with globalization—particularly neoliberalism—are dramatically transforming the context for both exposure and responses to extreme events. Neoliberalism entails a shift toward free-market approaches to management of international, national, and local economies. It is often portrayed as a monolithic exogenous force, yet in reality it transforms places in a differentiated, segmented, and highly uneven manner (Martin 2005). Some of the key components of neoliberal policy changes that differentially affect the urban context are (1) national policy changes that include opening markets to trade and capital investment; (2) structural adjustment and other fiscal programs that limit government spending and borrowing in cities; (3) privatization, including removal of the state from housing provision, and marketization of urban services; and (4) devolution of responsibility for urban governance to the local level (Brenner and Theodore 2002; Peck and Tickell 2002; Newman and Ashton 2004; Leichenko and Solecki 2005). This suite of changes, which is closely associated with new public management in social democratic states, also includes administrative reforms such as hands-on professional management, explicit standards for performance and accountability, a greater emphasis on output control, increased competition, contractualism, and application of private-sector management techniques (Christensen and Lægreid 2001).

New public management, as well as structural adjustment programs in many developing countries (which were part of earlier neoliberal reforms), have had severe effects on the urban poor, including on health, education, and provision of basic services (Walton and Seddon 1994; Mohan et al. 2000). These neoliberal changes have often meant a general reduction in the capacity of local governments to manage cities and to address the social service and welfare needs of poorer residents. Safety nets include unemployment insurance, social welfare programs, and health care services. In many cities there has been a marked decline in social supports for poor residents which, when combined with the privatization of many urban services and the establishment of fees for services, makes it difficult for many poor people to care for themselves or cope with misfortune. In many countries where government-supported safety nets are limited to begin with, most people rely heavily on extended families and communities, drawing upon social capital.

Other consequences of neoliberalism in cities include devolution and fragmentation of planning and governance, leading to a proliferation of unplanned and unregulated settlements. This fragmentation contributes to urban spatial expansion (and the attendant biophysical changes) and increased population densities in physically unsafe areas such as flood plains.[7] In many cases, new migrants and poorer residents concentrate in these hazardous areas because they are the only locations that are affordable (Davis 2006). Political fragmentation also means that disaster response efforts are more difficult to coordinate among local government agencies (Leichenko and Solecki 2006; 2008). During extreme events, there are fewer social services for people to turn to for assistance. After such events, the rebuilding and reconstruction are haphazard and depend mostly on private-sector actors.

Context Double Exposure in Urban Areas

While the preceding discussion emphasizes specific types of contextual changes in cities, it is critical to recognize that all of these changes—and others—are occurring together, either simultaneously or sequentially. The pathway of context double exposure explicitly focuses on these overlaps and points to ways that both global processes may increase exposure and decrease the capacity to respond to change, thus enhancing vulnerability of some individuals and groups while decreasing resilience of urban systems to future change.

There is clearly a close relationship between uneven outcomes from global change processes, as described in chapter 5, and increasing vulnerability. The growing wage and income inequalities in urban areas, for example, can be considered not only as outcomes of globalization but also as changes in the contextual environment for exposure and response to extreme events. However, the link between uneven outcomes and growing vulnerability is seldom direct or linear.[8] Present-day outcomes often lead to responses that influence the context in which *future* shocks and changes occur. The pathway of context double exposure looks at how the processes change the contextual environment and thereby influence the outcomes of future events.

The next section describes how both processes have transformed the contextual environment in and around New Orleans, increasing the vulnerability of many of the city's residents to Hurricane Katrina. Hurricane Katrina brought to light the existing poverty and inequality of New Orleans, the consequences of coastal environmental degradation, and the institutional weaknesses within government agencies. Context double exposure provides a framework for understanding how the processes of global environmental change and globalization influenced these conditions and increased vulnerability to the hurricane. It also shows the important role of the contextual environment in the recovery and reconstruction phase, which may either enhance or reduce the city's resilience to all types of future change.

Signatures of Urban Disaster: New Orleans and Hurricane Katrina

Hurricane Katrina hit land in southeastern Louisiana on August 29, 2005. Direct exposure to the hurricane resulted in considerable damage throughout the U.S. Gulf Coast region, including portions of Louisiana, Mississippi, and Alabama. Yet within the port city of New Orleans, which was narrowly spared a direct hit by the hurricane, immediate damage was minimal. For a very short time many of the city's residents believed that the so-called Crescent City had "dodged the bullet." However, breaches in the city's protective levees soon became apparent, and subsequent flooding caused significant loss of life and tremendous physical damage. By the time the flood-waters had receded, Katrina had become the costliest weather-related disaster in U.S. history, responsible for a loss of roughly 1,800 human lives, as many as one million displaced people, and over $80 billion in damage to homes, businesses, and public infrastructure.

While almost all the residents of the New Orleans metropolitan area were affected by the storm, a disproportionate share of those who died were members of underprivileged or marginalized social groups (Tierney 2006; Gullette 2006). The elderly were especially vulnerable to Katrina. Estimates are that more than 60 percent of those who died were over the age of 60 (Gullette 2006). While many died in nursing homes or institutions that were responsible for their care, some of the city's older residents perished in their own homes because of illness, weakness, and lack of mobility. Another highly vulnerable group was poor African-Americans. Many of the worst-affected neighborhoods, such as the lower Ninth Ward, were home primarily to lower-income African-Americans, many of whom were trapped after the hurricane with no access to transportation for evacuation.

In addition to loss of life and destruction of property, Hurricane Katrina also had widespread economic effects, including increases in energy costs and basic commodity prices, disruptions in shipping and trade along the Mississippi River, and adjustments to national and international travel and tourism (Reeves 2005). Longer-term effects were felt in all parts of the United States, particularly the cities and towns where Katrina victims took refuge. Cities such as Houston and Atlanta absorbed thousands of refugees, a situation that put demands on schools, health facilities, and social services, as well as communities in these areas that were faced with the task of absorbing many newcomers on short notice.

Within the media and popular literature, much attention has centered on the immediate causal factors that could explain "what went wrong" and who was to blame in New Orleans (Young 2006). These accounts focus on failed levees and the role of the U.S. Army Corps of Engineers; the Federal Emergency Management Agency (FEMA) and its lack of preparedness and inability to evacuate the population from the city; climate change and its role in extreme events; business and oil interests and their role in the destruction of marshes and wetlands; societal racism that placed poor African-American communities in the lowest-lying areas and delayed rescue efforts; and a reluctance or inability of some to leave the city before the hurricane struck (Van Heerden and Bryan 2006; Dyson 2006; Brinkley 2006; Horne 2006). While there is some validity to each of these explanations, the double exposure framework shows how global environmental change and globalization contributed to the context for disaster in New Orleans.

A Changing Context in New Orleans

Hurricanes have long been recognized as a major threat to New Orleans. The city's vulnerability is based on the convergence of several factors: the fact that 80 percent of the city lies below sea level, relying on levees and other human-made infrastructure for flood protection; high degrees of income inequality and racial segregation; and relatively weak public institutions. These local contextual conditions—all of which have been identified and widely discussed by urban geographers and disaster researchers (e.g., Cutter 2006; Cutter et al. 2006; Colton 2005)—are certainly central in explaining the physical devastation and uneven social outcomes of Katrina. Less well-recognized is the role that processes of global environmental change and globalization played in exacerbating the region's hurricane vulnerability. Drawing from the preceding discussion of contextual changes in urban areas, we will show how global change-related transformations of the biophysical, economic, and institutional context in the region affected both exposure and capacity to respond to Katrina (see figure 6.1).

Biophysical Context

Humans have been changing the biophysical context in the Mississippi Delta region for centuries (Colton 2000). Although major environmental transformations have

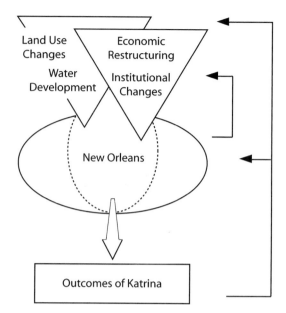

Figure 6.1
Context double exposure in
New Orleans: Biophysical,
institutional, and economic
changes altered the contextual
environment prior to Hurricane
Katrina.

occurred in nearly all large cities, New Orleans is unusual in terms of the scope of
what has been done to ensure viability for human settlement. These transformations, which include construction of levees, drainage of wetlands, and dredging
of rivers, have made New Orleans habitable for a population of over half a million
people. At the same time, these changes have made the region "the poster child
for problems threatening the world's deltas, coastal wetlands and cities on the sea"
(Fischetti 2001, 78).

The biophysical changes in New Orleans and its surroundings illustrate how
local-scale environmental changes can lead to larger-scale environmental transformations. In the case of New Orleans, subsidence and loss of protective barrier
islands are the cumulative result of changes that have been occurring throughout
the Mississippi River basin over the past century but have accelerated in recent
decades (Davis 2000). Construction of dams, reservoirs, and levees to protect riv-
erine communities and industries along the entire length of the Mississippi has
affected the sediment deposits necessary to maintain and replenish the delta and
barrier inlands (Sparks 2006). As a result, sediment carried to the mouth of the
Mississippi no longer builds up in the delta and on the eroded shorelines of the barrier islands but instead is deposited directly into the ocean (Fischetti 2001). Without
this renewal process, the compression of sands and silts creates subsidence as well
as the loss of marshes and wetlands in the region. Recent estimates suggest that the
rate of subsidence in New Orleans is accelerating, with as much as one meter loss
occurring between 2002 and 2005 (Dixon et al. 2006).

Biophysical changes in the region are also linked to economic activities, particularly expansion of the region's facilities for recovery, and transport of oil and

gas (Turner 2004). For example, the Mississippi River Gulf Outlet (MRGO), a 76-mile (122 km) channel developed in 1965 to shorten shipping times between New Orleans and the Gulf of Mexico, has been implicated in the destruction of more than 20,000 acres of adjacent wetlands (U.S. Army Corps of Engineers 1999).[9] Other channels and canals that were cut through the marshes to transport fossil fuels have also significantly damaged the area's marshes and wetlands. As was noted by Fischetti (2001, 79), "The network of canals ... gives saltwater easy access to the interior marshes, raising their salinity and killing the grasses and bottomwood forests from the roots up. No vegetation is left to prevent wind and water from wearing the marshes away." The subsidence of land and disappearance of the delta and barrier islands that buffer New Orleans from the Gulf of Mexico, in combination with the cutting of new channels through marshes, means that storm surges have a direct path into the city (Fischetti 2001).

The biophysical vulnerability of the Mississippi River Delta region was clearly revealed by Hurricane Katrina, and continued environmental changes associated with sea level rise are likely to increase exposure to future hurricanes and extreme events. As was emphasized in a 1998 report on Louisiana's coastal land loss:

> If recent loss rates continue into the future, even taking into account current restoration efforts, then by 2050 coastal Louisiana will lose more than 630,000 additional acres of coastal marshes, swamps, and islands. The loss could be greater, especially if worst-case scenario projections of sea-level rise are realized, but in some places there is nothing left to lose. (Coast 2050, 1998, 1).

As was noted earlier, sea levels are expected to rise significantly in the next 100 years because of climate change-related thermal expansion of the oceans and melting of glaciers and ice sheets (Church et al. 2001; Meehl et al. 2005; Alley et al. 2005). Climate change is also likely to raise sea surface temperatures, a factor that may increase the strength of hurricanes (Trenberth 2005; Emanuel 2005).

Socioeconomic Context

In addition to biophysical factors, the vulnerability of many New Orleans residents also stems from socioeconomic conditions, including high rates of unemployment, low levels of educational attainment, high percentages of female-headed households, and high degrees of racial segregation (Glasmeier 2005; Hartman and Squires 2006). In 2005 nearly 33 percent of the city's families with children lived below the U.S. poverty line, a rate more than double the U.S. average (U.S. Census Bureau, 2005). Using metrics of persistent poverty and social vulnerability, respectively, Glasmeier (2005) and Cutter et al. (2006) each demonstrates that these conditions have been persistent in the city for more than four decades.[10] New Orleans Parish, in particular, ranked in the top 3 percent of the most socially vulnerable counties in the United States in both 1960 and 2000 (Cutter et al. 2006).

Although high degrees of poverty and other social problems influenced vulnerability to Katrina in New Orleans, globalization-related economic pressures have exacerbated these contextual conditions. Like many cities in the United States and other advanced economies, New Orleans has experienced a large-scale reduction in industrial employment (Whelan 2006). In response, the city has sought to remake itself as a site of global consumption, with particular emphasis on national and international tourism via the "marketing of Mardi Gras" (Gotham 2002). While sales taxes associated with tourism provide the city with a large share of its tax revenue, the industry is characterized by low-wage jobs, many of which offer no benefits.[11] Prior to Katrina, the entertainment, food, arts, and accommodation sectors together were the city's second largest employer, accounting for nearly 16 percent of the city's jobs. In contrast, for the United States as a whole in 2005, these sectors accounted for approximately 8 percent of jobs.[12]

Also because of specialization in tourism-related industries, there were few livelihood options for displaced workers when tourism was disrupted by the storm. Furthermore, because reconstruction efforts following the disaster have focused largely on the city's historic areas and its gambling casinos, the city is likely to become even more dependent on tourism in the future (Cutter et al. 2006). This emphasis on reconstruction of tourist spaces is consistent with the idea of cities as sites of consumption (Miles and Miles 2004), yet it also reflects the broader influence of neoliberalism on reconstruction efforts, particularly a growing reliance on private-sector actors. Given that rebuilding efforts have been dominated by private actors, it is not surprising that repair efforts have focused largely on for-profit enterprises such as hotels and tourist areas while neglecting schools and other public facilities. By the end of 2006 approximately 90 percent of the city's hotels were fully operational (Liu et al. 2006) while only 74 percent of its public schools had reopened. Within the hardest hit parishes these differences are even more striking. In Orleans and St. Bernard Parishes, only 49 and 20 percent of public schools, respectively, had reopened by late 2006 (Liu et al. 2006).

Reconstruction or rehabilitation of damaged housing—which has been left largely to the private sector—has also proceeded at a slow pace. For the city as a whole, the hurricane damaged an estimated 160,000 housing units. As of December 2006 only 47,000 permits had been issued for housing repairs or renovations (Liu et al. 2006). Reconstruction efforts in areas dominated by low-income rental housing or public housing, such as the Lower Ninth Ward and the Calloipe, Iberville, and St. Bernard Area Projects, have lagged even further behind, such that many of these areas remain uninhabitable (Center for Social Inclusion 2006; Crowley 2006). This lack of housing has been a key stumbling block for the recovery efforts (Whelan 2006; Liu et al. 2006). At the end of 2006 the city had a substantially lower population, and a different social and economic structure, than before Katrina, with fewer poor residents and fewer African-Americans (Logan 2006). The population of Orleans Parish as of November 2006 was estimated at approximately 190,000, more than 55 percent less than its population of 444,000 in the year prior to Katrina (Louisiana Department of Health and Hospitals 2006). The city's future racial makeup is also likely to be substantially different, particularly if rebuilding of heavily damaged areas in lower-income neighborhoods does not proceed (Logan 2006).

Institutional Context

The effects of neoliberal institutional changes can be seen in many facets of the city's response to Hurricane Katrina. A weakened and ineffective Federal Emergency Management Agency (FEMA), poor coordination of local and state government efforts, lack of appropriate maintenance of the levees, and the general lack of disaster preparedness all reflect the retreat of the state from active responsibility for disaster management.[13] The implications of a neoliberal weakening of public institutions in a city with high levels of concentrated poverty were apparent in the immediate aftermath of Katrina. Those with the fewest resources responded to the flooding by congregating at the New Orleans Convention Center and the Superdome with the expectation that FEMA or local and state governmental agencies would help them. In many cases this help took days to arrive. Patterns of evacuations of hospitals in New Orleans reflected a neoliberal approach whereby those patients who had the financial resources and private insurance necessary to use private hospitals were evacuated in a timely manner. In some cases, those in public hospitals, which serve primarily the poor, had to wait up to five days after the hurricane to be evacuated.[14]

In addition to a lack of disaster preparedness, the weakening of state institutions—particularly reduction of social and mental health services—is also likely to affect the long-term capacity of individuals to recover from Katrina. Beyond the immediate loss of life, the associated trauma is having long-term effects on the health and well-being of the survivors. Recovery from an event such as Katrina depends largely on the ability to recover the sense of security that was lost by those, particularly the elderly, who were displaced from their homes (O'Brien et al. 2005). Post-traumatic stress disorders and other mental health effects often continue long after an event, making individuals less able to cope with future events (Norris 2005). Inadequate social services and mental health support for victims are likely to exacerbate these conditions.

Future Trajectories

Hurricane Katrina hit upon a contextual environment in 2005 that had already been transformed by both global environmental change and globalization, contributing to a disaster of unprecedented significance in the United States. In terms of the city's future, the outcomes and responses to the hurricane will, in and of themselves, become critical factors in shaping the contextual environment. For example, much of the cleanup and reconstruction work in New Orleans has been undertaken by private contractors who hire informal labor. Consequently, there has been a large-scale in-migration of Latin American workers to this region, many of whom send money back to their home countries. Many of these transnational migrants, who are representative of a growing component of globalization (Smith 2001), may ultimately remain in the region. Responses to wetland loss, management of the Mississippi River flows, and the future of the MRGO will also have profound implications for both the city and the surrounding region. The city's future contextual environment and its resilience will depend to a large extent on the decisions made and actions taken in the wake of Katrina.

Conclusion

Within this century, cities will become the most important form of social organization for the majority of the world's population. Extreme or extraordinary events in cities often play out very differently according to the context, in some cases resulting in only minor disturbances and in others cases resulting in disasters or catastrophes. Often such events reveal underlying vulnerabilities that are less visible, or are even hidden, during "normal" times. The example of Hurricane Katrina clearly revealed an underlying context of vulnerability in New Orleans, yet it also demonstrated how global change processes affected this context. The disastrous outcomes of Hurricane Katrina not only were the result of the city's geographic location, failed levees, or history of poverty and inequality, but also reflected larger-scale biophysical, economic, and institutional transformations. Thus the devastation from Katrina is not likely to be a unique event. Rather, Katrina may be a signature of disasters to come as global change processes increase urban vulnerabilities to all types of extreme events.

Context double exposure shows how both global environmental change and globalization can undermine human security in urban areas. The increased likelihood of shocks related to both processes places a high premium on increasing the resilience of cities. Building resilience in cities can help address the new threats, including extreme climate events, sea level rise, financial disruptions, and disinvestment. As the situation is described by Berkes (2007, 284):

> Resilience puts the emphasis on the ability of a system to deal with a hazard.
> It allows for the multiple ways in which a response may occur, including
> the ability of the system to absorb the disturbance, or to learn from it and to
> adapt to it, or to reorganize following the impact.

By recognizing how both global environmental change and globalization together affect urban contexts, we can begin to identify actions and policy responses that enhance the capacity of urban systems and urban residents to adapt to all types of future shocks and stresses.

Dynamic Feedbacks and Accelerating Changes

7

We tend to think of the different challenges we face in a piecemeal
fashion. Poverty, disease, climate change, jobs, biodiversity loss,
social justice, pollution, education, fisheries collapses, invasive
species, national security—these are all viewed as separate issues....
This fragmented approach is myopic and unfortunate, especially
because there are subtle but demonstrable connections between
and among many of the environmental and socioeconomic ones.
Moreover, the fragmentation may inadvertently generate tsunami-like
waves.
—Jane Lubchenko, "Waves of the Future: Sea Changes for a
Sustainable World"

The notions of increasing connectivity and accelerating change are frequently used
to characterize both global environmental change and globalization. Global environmental change discourses stress the connections and dynamic feedbacks among
physical, ecological, and social systems, all of which are thought to form part of a
broader and interconnected Earth System (Steffen et al. 2004). Globalization discourses emphasize the growing spatial and temporal interconnectedness of economic, political, social, and cultural systems, and they stress the influences of
space-time compression and a "speeding up" of all facets of communication and
interaction (Harvey 1990; Held and McGrew 2000; Homer-Dixon 2006; Barnett
et al. 2007).[1] Less well-recognized within the various discourses are the linkages
and feedbacks *between* processes of global environmental change and globalization.
This chapter explores these issues, showing how linkages and feedbacks between
the two processes may both enhance global connectivities and contribute to accelerating rates of global change.

This chapter first considers how the different global change discourses interpret the linkages between the two processes. Most research in this area emphasizes globalization as a driver of environmental pollution and degradation. Relatively few studies recognize that global environmental change may contribute to globalization, and there is virtually no recognition of iterative feedbacks between the two processes. Next is presented a case study that draws on the pathway of feedback double exposure to show how the two processes are dynamically linked in the Arctic region. The Arctic is changing rapidly as a result of both climate change and the expansion of international shipping and transport. Direct linkages between the processes emerge because reductions in sea ice due to climate change open new shipping routes and provide greater access to oil and gas reserves of the region. Feedbacks may occur, as exploitation of the region's oil and gas resources leads to further increases in greenhouse emissions and a further expansion of international shipping and trade.

In highlighting these linkages, feedback double exposure reveals how global change processes can be mutually reinforcing, thus posing challenges to sustainability. In the case of the Arctic, feedbacks between climate change and globalization may lead to an enhanced version of what Beck terms "the boomerang effect," whereby risks eventually come back to affect those who produce or profit from them (Beck 1992). Economic opportunities associated with an open Arctic Ocean are expected to be a boon to Arctic countries and to international trade as a whole, benefiting transport and resource extractive industries, including oil and gas. Over time, however, as global environmental change and globalization processes mutually reinforce each other, coastal regions around the world—including those in the Arctic—will face serious challenges associated with higher sea levels, increased storms surges, and changing weather patterns, not to mention changes in biodiversity and ecosystem services. The Arctic case study demonstrates that in an increasingly connected world, outcomes that accelerate either process of global change are likely to have widespread ramifications and long-term consequences.

Linking Global Environmental Change and Globalization

Among the global change discourses, globalization is widely regarded as a major "cause" of environmental change. Although the biophysical discourse increasingly recognizes that there may be both positive and negative consequences of globalization, most research focuses on the negative implications for water, the atmosphere, forests, and agriculture (Steffen et al. 2004). The human-environment discourse, in contrast, looks at the consequences of globalization for the structural characteristics of social-ecological systems at various scales, highlighting changes in connectedness, speed, and scale (Young et al. 2006). Such research emphasizes the ways that globalization influences the resilience, vulnerability, and adaptability of these systems. The critical discourse on global environmental change, as well as the malevolent discourse on globalization, stresses the negative environmental consequences of globalization, with some of the most potent forces being increased trade and transport, growing urban consumption, and changes in institutional structures for resource management (Speth 2003; Roberts and Thanos 2003; McCarthy 2004). Within these discourses, it is frequently argued that environmental and social problems are inherent in the prevailing model of globalization, with its emphasis on privatization, enclosure, and commodification (Mander 2003).

In contrast to seeing globalization as a cause of environmental degradation, some global change discourses emphasize the potential for globalization to improve environmental conditions. The benign globalization discourse draws attention to technological improvements and market-oriented approaches to international environmental cooperation and management as a means of successfully addressing environmental change (Marcotullio and Lee 2003; Nepstad et al. 2006). Drawing upon concepts of ecological modernization and the environmental Kuznets curve, this research suggests that globalization-induced economic development can bring about improvements in environmental quality via the transfer of environmentally friendly production technologies, demands by higher-income citizens for cleaner local environments, and increases in the capacity of local governments to control environmental pollution (Stern et al. 1996).[2]

The transformative discourse takes a slightly different view, seeing the potential for changes in global economic, political, and institutional systems as ways to address global environmental issues such as climate change (Held et al. 1999; Sklair 2002; Bulkeley and Betsill 2003). Indeed, new institutional forms under globalization are increasingly recognized as a critical factor in environmental management at all levels (Brewer 2003; McCarthy 2004; Lemos and Agrawal 2006; Mol 2006; Liverman and Vilas 2006; Zimmerer 2006). It has also been suggested that international scientific and political networks, international environmental agreements, and international political and social movements can provide new ways for the global community to respond collectively to the challenges of environmental change (Mittelman 2000; McDonald 2006; IPCC 2007b).

Drawing from these debates, a large body of empirical research has explored the environmental consequences of globalization. Topics of these studies include the effects of foreign direct investment on pollution levels in host countries, the impacts of globalizing food consumption preferences on ocean species such as tuna and shrimp, the connections between expansion of international airline transport and greenhouse gas emissions, and the influence of changing institutions under globalization on ecosystem resilience and rural poverty (Johnson and Beaulieu 1996; OECD 1997; U.S. GAO 2000; Tom Jones 2002; Eskeland and Harrison 2003; Aggarwal 2006; Jorgensen 2007). Other studies explore the possibility that globalization can exert downward pressures on national environmental standards, creating a "race to the bottom" (Esty 1994; Bredahl et al. 1996; Rauscher 1997; Porter 1999; Schofer and Granados 2006). The findings of these empirical studies are generally mixed. In a review of recent literature on neoliberal policies and local environments in Latin America, for example, Liverman and Vilas (2006) found that the impacts varied by nation and place, depending on the contextual conditions. Despite the negative picture painted by a number of studies, some places and people have adapted well to and benefited from neoliberalism (Liverman and Vilas 2006).

While a large body of literature emphasizes the role of globalization as a force for environmental change, global environmental change is not typically seen as a driver of globalization processes. A few exceptions include studies of how climate change might affect agricultural trade patterns and those linking climate change to the expansion of transportation routes (Reilly et al. 1994; ACIA 2004). Rather, environmental change is more often seen a threat to the global economy. Land degradation and climate change, for example, are recognized as factors that may limit prospects for competitive participation in the global economy in some regions, particularly sub-Saharan Africa. There is also growing recognition that climate change may impede economic growth in developed countries (Stern 2006). Indeed, dire warnings of how environmental degradation may harm prospects for development in the future have long been a fixture within the sustainability literature (WCED 1987).

Although many of the global change discourses acknowledge direct causal linkages between globalization and global environmental change, the pathway of feedback double exposure emphasizes the potential for indirect linkages and feedbacks. As was mentioned in earlier chapters, responses to each process can contribute to the drivers of the other process. For example, rural-urban migration in Latin America, which is in part a response to the cultural and economic attraction of urban life, has contributed to the abandonment of grazing and agricultural lands and the recovery of forest ecosystems as well as to rapid urban expansion (see Aide and Grau 2004; Hecht et al. 2006). The expansion of urban areas, however, creates significant land use changes that can influence ecosystem services, biodiversity, and the local and regional climate (Solecki and Leichenko 2006; Foley et al. 2005). Feedback double exposure shows that the two processes are dynamically intertwined, such that they not only increase global connectivities but also accelerate processes of global change.

The example that follows draws attention to the direct and indirect linkages between global environmental change and globalization in the Arctic, as well as the feedbacks that may result from these linkages. In the Arctic the linkages between the two processes are highly visible, but the feedbacks have not been clearly articulated. Emphasizing the dynamic nature of the two processes shows how the linkages can generate feedbacks that influence the direction, magnitude, and rate of change.

Sea Ice, Shipping, and Oil Extraction: Winners and Losers in the Arctic

Processes of global environmental change and globalization are becoming increasingly visible in the Arctic, where activities originating in other parts of the world are causing profound environmental and social transformations. Heavy metals and persistent organic pollutants (POPs) used in industrial and agricultural processes have polluted the Arctic environment, accumulating in species at the top of the food chain (AMAP 2005). The combustion of fossil fuels throughout the world is increasing atmospheric concentrations of greenhouse gases, leading to climate change and a rapid melting of Arctic permafrost and sea ice (ACIA 2004). At the same time, a growing global demand for hydrocarbons and mineral resources, particularly in Asia, supports the expansion of oil, gas, and mineral exploration and extraction in the Arctic region. Expansion of commercial fishing and a growing international tourist presence in Arctic areas are also placing increasing pressure on Arctic communities and ecosystems. Closely related to these economic activities, there has been a dramatic increase in ship traffic in the Arctic, which has raised a number of environmental concerns regarding water and air pollution, the introduction of exotic species, and the potential risk of oil spills (PAME 2000).

Arctic residents have a long history of adapting to changing conditions (Crate 2006). Nevertheless, today they face an unprecedented combination of rapid and stressful changes related to global environmental change, globalization, and other processes (AHDR 2004). The dramatic changes taking place in the Arctic also have implications that extend beyond Arctic communities, not only because southern power centers have economic and strategic interests in the Arctic, but also because Arctic communities are gradually expanding their reach south (Csonka and Schweitzer 2004). The significance of the Arctic cannot be understated, not the least because climate change in the region is considered a harbinger of much larger changes that will occur around the globe (Chapin et al. 2004; ACIA 2004; Yardley 2007).

One change that has drawn particular attention in recent years is the reduction of Arctic sea ice (Johannessen et al. 1999; Comiso 2002; Laxon et al. 2003; ACIA 2004; Laidler 2006; IPCC 2007b). Loss of sea ice is changing the biophysical context for heat and mass exchanges and for ocean stratification, and it is also creating a new biophysical context for economic activities. This new context is likely to open additional shipping routes through the Arctic, leading to expanded global transport and an acceleration of global trade flows. The next two sections will review the bodies of evidence associated with climate change and transport expansion in the Arctic, emphasizing the direct causal linkages between reduction of sea ice and growth of shipping that are recognized within these literatures. We will then draw attention to potential feedbacks, showing how actions taken in response to changing conditions in the Arctic may ultimately accelerate the process of climate change (see fig. 7.1).

Climate Change in the Arctic

Climate change is clearly evident in the Arctic region. Since the late 1960s, higher average temperatures, melting of permafrost, and reductions in the extent and duration of snow and sea ice have been documented in areas throughout the region (ACIA 2004; IPCC 2007a; IPCC 2007c). Long-time residents of Arctic communities confirm that there have been dramatic changes over the past several decades in seasonality and thickness of sea ice coverage, suitability of snow for igloo construction, and patterns of species migration (Krupnik and Jolly 2002; ACIA 2004; Laidler 2006). While the Arctic is particularly prone to climatic fluctuations and

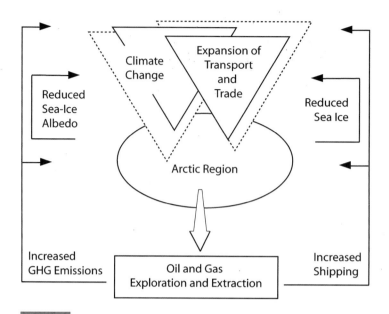

Figure 7.1
Feedback double exposure in the Arctic. Climate change and the
expansion of transport in the Arctic each create feedbacks which
contribute to processes of global change.

has undergone many climatic shifts in past millennia, the current changes have
been attributed primarily to human activities associated with the emission of greenhouse gases (Holland 2002; ACIA 2004). Concerning future trends, climate models
predict significant changes in the Arctic, and particularly in sea ice cover over the
next 50 years (Holland et al. 2006).

Climate change within the Arctic also tends to be self-reinforcing (ACIA 2004).
The melting of sea ice reduces the surface reflectivity, or albedo. Darker surfaces
absorb more solar radiation and reflect less longwave radiation back into space,
resulting in a warming effect that accelerates the melting of sea ice (Winton 2006).
This feedback not only will enhance the rate of Arctic melting, but may also push
the region toward a "tipping point" whereby an abrupt elimination of Arctic sea
ice leads to regional temperature increases above that expected by warming of the
surrounding region (Winton 2006; Holland et al. 2006). Another type of feedback
may occur as a result of melting of permafrost and the release of stored carbon and
methane. Because permafrost currently acts as a significant global sink, its melting
may transform the Arctic region into a net source of greenhouse gases, leading to
further global warming (Stokstad et al. 2004; Smith et al. 2004).

The melting of the sea ice and loss of permafrost represent dramatic changes in
biophysical conditions of the Arctic region that challenge the resilience of both people and ecosystems. For example, communities and groups that depend upon snow,
ice, and associated flora and fauna for their livelihoods (e.g., reindeer herders) are

likely to experience negative outcomes from climate change (ACIA 2004). Although dramatic changes associated with Arctic warming represent stressors for some, they present opportunities for others. Shipping companies, for example, may view climate change as a welcome opportunity to increase transport across the Northern Sea Route (NSR), as ice-free conditions make this corridor cheaper and easier to navigate. It may also open up new oil, gas, and mineral resources for exploitation, a situation that is likely to have both positive and negative consequences for residents of the Arctic.

The opportunities for expanded transport and shipping presented by reduced sea ice are recognized in the climate change literature, including in the IPCC Third Assessment Report (TAR) and the Arctic Climate Impact Assessment (ACIA):

> The impact of climate warming on marine systems is predicted to lead to loss of sea ice and opening of sea routes such as the Northeast and Northwest passages. Ships will be able to use these routes without strengthened hulls. There will be new opportunities for shipping associated with move-ment of resources (oil, gas, minerals, timber), freight, and people (tourists). (Anisimov and Fitzharris, 2001, 828).

> The Russian Arctic holds significant reserves of oil, natural gas, timber, cop-per, nickel, and other resources that may best be exported by sea. Regional as well as trans-Arctic shipping along the NSR is very likely to benefit from a continuing reduction in sea ice and lengthening navigation seasons (ACIA 2004, 83).

Nevertheless, the potential for feedbacks between increased transport and trade, Arctic resource exploitation, and climate change has not been well recog-nized in the climate change literature. This is similarly true in the literature on shipping and transport through the Arctic. As will be shown in the next section, although much has been written about shipping in the Arctic, particularly across the Northern Sea Route, there has been relatively little attention given to the impli-cations of these developments for the global environment.

The Northern Sea Route and Global Trade

In recent years, shipping activities have increased significantly in the Arctic region, with Russia representing the region's largest user of water transport. Transport cur-rently plays an important role in the Arctic economy, contributing from 5 to 12 percent of the production value, depending on the region (Duhaime 2004). The transport sector links the Arctic to the rest of the world, exporting local products and importing goods on which residents of Northern countries depend (Duhaime 2004). Transport vessels include cargo ships, fishing boats, cruise ships, and research ships. In addition, icebreakers, tugs, and the transport of vessels for scrap-ping form part of the shipping traffic in the Arctic (PAME 2000).

The Northern Sea Route (NSR) exemplifies many of the opportunities and chal-lenges associated with shipping and transport in the Arctic region.[3] The NSR is, according to the official Russian definition, a collection of sailing lanes that extend from Novaya Zemlya in the west to the Bering Strait in the east. It includes the main part of the stretch known as the Northeast Passage, which connects the Atlantic and Pacific Oceans along the northern coast of Eurasia (Østreng 1999). The NSR covers between 2,200 and 2,900 nautical miles of water, most of which is covered with both seasonal and year-round ice. The route is navigable throughout the year in the western part, but along the central and eastern coast of Eurasia it is passable only from mid-July to mid-September, when much of the ice has melted (Østreng 1999).[4]

Historically, the Northern Sea Route has played an important role in economic development of the Russian Arctic. In combination with numerous rivers flowing into the Arctic Ocean, the NSR forms the water-transport system that handles most of the cargo going into and out of the region. The growth of cargo shipment along the NSR since 1960 has been closely linked to the exploitation of the region's diverse and rich natural resources, which include petroleum, natural gas, nickel, tin, lead concentrate, titanium concentrate, apatite concentrate, diamonds, gold, and platinum (Granberg 1998; Khvochtchinski and Batskikh 1998).

Since the early 1990s, international efforts have been made to make the NSR a viable shipping route, connecting distant markets along the Atlantic Ocean with those along the Pacific (Ragner 2000).[5] The NSR can dramatically reduce the transport distance between Europe and northeast Asia, capturing some of the trade that is currently directed through the Suez and Panama Canals. This opening can potentially cut travel time for shipping and also increase overall levels of international trade. For example, using the NSR can take as little as 10 to 15 days for shipments between Japanese and North European ports, compared with 30 to 33 days through the Suez Canal (Østreng 1999). This rerouting is important for countries that ship most of their exports by water, including China. The commodities that may benefit most from the NSR include oil, gas, and mineral resources—many of the resources that are abundant in the Arctic and considered essential to fuel economic growth in Asia, particularly in China.

While some of the factors limiting expansion of the NSR are financial and political, ice cover also represents a major limiting factor for commercial navigation and transport.[6] Along the NSR, ice conditions in the East Siberian and Chukchi Seas present the most formidable challenges to passage, with the former experiencing the least summer ice-melt of any region along the NSR (Brigham et al. 1999).[7] These combine with meteorological factors (wind, visibility) and oceanographic factors (waves, currents, sea level) to limit the navigation season to a period from June to October (Proshutinsky et al. 1998). The areas with the most difficult ice conditions are also the most shallow, presenting additional challenges to navigation (Østreng 1999). Consequently, the main factors in selecting transport routes along the NSR are the nature and distribution of sea ice and the presence of sufficiently deep water (Brigham et al. 1999).[8]

Like the climate change literature, the Northern Sea Route literature recognizes the potential benefits of reduced sea ice to facilitate transport. Brigham et al. (1999) notes, for example, that more favorable ice conditions cause less damage to ship hulls due to less collision with ice, and a smaller accompanying icebreaker fleet is required; both of these advantages will reduce shipping costs. Similarly, Østreng

(1999) comments that transport time can be reduced significantly in the absence of sea ice. Concerning the environmental impacts of increased shipping, the transport literature generally focuses on local environmental hazards associated with expanded shipping, including oil spills and other types of discharges, point source pollution, and noise (Moe and Semanov 1999). The broader environmental consequences of opening the Arctic, however, and particularly the connections to future climate change, are only beginning to be recognized.

Feedbacks

As was discussed earlier, the possibility of accelerated climate change due to a sea-ice/albedo feedback is well documented within the scientific literature (Winton 2006; Holland et al. 2006). Double exposure to reduced sea ice and transport expansion sets up the potential for another type of climate change feedback, whereby both processes together facilitate increased fossil fuel extraction, increased international transport and trade, and increased net emissions of greenhouse gases, and hence lead to further climate change.

The Arctic possesses a rich abundance of natural resources, including substantial oil and gas resources in the Barents, Kara, Laptev, East-Siberian, and Chukchi seas. In fact, more than 95 percent of the Russian potential offshore hydrocarbon resources are concentrated in the Arctic Ocean (Nikitin and Mirzoev 1998). Consequently, much of the international shipping traffic in the Barents Sea and northern part of the Norwegian Sea is dominated by vessels going to and from Russian ports, such as Murmansk, Arkhangels, and Kandhalaksha, in connection with oil exports from Russia (PAME 2000). Under current plans, the number of tankers transporting oil from Russian ports through the Barents Sea is expected to rise from 166 in 2002 to 650 by 2015, and most of the increase is expected to be tankers larger than 100,000 dwt, compared with present sizes of 30,000 dwt and smaller (DNV 2003).

The oil and gas resources in the Barents and Kara Sea regions are already being explored and exploited. These western regions are technically more accessible and economically more profitable than the eastern regions, which have more severe hydrographic climate conditions (Nikitin and Mirzoev 1998). Beyond existing activities, recognition of the emerging opportunities in the Arctic has prompted a new interest in developing the energy resources of the region. These plans are exemplified by the Norwegian government's *nordområdestrategi*, a strategy to develop oil and gas resources in the Barents Sea and promote development (including tourism and fisheries) in northern Norway (Norwegian Ministry of Foreign Affairs 2006). Such plans for development and economic integration of the Arctic not only are regionally based but also are fueled by institutional reforms and international collaborations linked to globalization, which are facilitating both resource extraction efforts and environmental initiatives (Holland 2002; Stokke and Hønneland 2007). As Heininen (2004, 221–222) points out:

> Northern economies are increasingly integrated into the globalized world
> economy and the importance of northern regions may grow with the

increased demand for strategic minerals and oil and gas, with larger companies with more capital taking an interest in the region, and with technology creating easier access to raw material sources. This integration is driven more by major states and transnational corporations than by regional actors.

The further development of the Northern Sea Route, as well as the development of other transportation routes such as the Northern Maritime Corridor,[9] will play an important role in facilitating the extraction and export of energy resources from the Arctic. Yet changes in ice conditions will also be critical. In fact, the defining factor in terms of organizing surveys, prospecting, and production works for oil and gas in the Arctic are ice regime characteristics, such as the presence of ice, ice types, and the geometric and kinematic characteristics of drifting ice formations (Nikitin and Mirzoev 1998). Reduced ice conditions as the result of climate warming are thus likely to have an enormous impact on both resource extraction and international trade by facilitating oil and gas exploration and production, increasing the length of the shipping season, reducing the demand for expensive icebreakers, and eventually opening additional year-round, ice-free shipping lanes.

The Arctic energy reserves that will be made accessible via reduced sea ice and expanding shipping networks will add to the world's available supply of oil and gas. This represents a potentially significant feedback to climate change, particularly since hydrocarbons in the Arctic are presently estimated at as much as 25 percent of the world's remaining oil and gas resources (Arctic Energy Summit 2006). Rather than serving as a substitute for current sources of oil and gas, new reserves are likely to supplement existing supplies, potentially continuing current trends in global consumption of fossil fuels. On the basis of existing technologies and consumption patterns, global energy consumption is projected to increase by 53 percent from 2006 to 2030, with fossil fuels (and especially oil) supplying much of this new energy (International Energy Agency 2006). According to these trends, global emissions of carbon dioxide are expected to reach 43 billion metric tons by 2030, an increase of approximately 55 percent from 2006 levels (Energy Information Administration 2006). This "business as usual" scenario—which projects substantial increases in oil consumption in China and India and continued high levels of consumption in the United States—appears all the more likely if there is a large influx of new supplies from the Arctic. Moreover, a tightening of global oil supplies and increasing energy prices make it increasingly profitable to extract oil from harsh Arctic environments (Chapin et al. 2004). Growing oil consumption from the Arctic and other sources is thus expected to substantially increase global carbon dioxide emissions and is thereby likely to accelerate climate change.

New advances in carbon capture and storage technologies may be used in energy production in the Arctic, and eventually decarbonization technologies may allow fossil fuels with low emissions of greenhouse gases to be used in distributed sources (such as automobiles) (Metz et al. 2005). However, these technologies are not expected to be available until after 2020, at which time emissions—and the impacts of climate change—will already have increased significantly (Greenpeace 2007). Furthermore, there is little understanding of the ecological impacts of carbon capture and storage, particularly for deep water/soil ecosystems (Johnston and Santillo 2002). In the meantime, "we are entering into the unknown with our climate" (Karl and Trenberth 2003, 1722). The linkages and feedbacks between reduced climate change, sea ice, oil and gas exploitation, and shipping and trade are likely to accelerate rather than mitigate climate change.

Uneven Outcomes

The vision of a warmer Arctic, bustling with economic activity, represents paradise for some but a catastrophe for others. Indeed, the pursuit of new opportunities in the Arctic strongly aligns with the interests and values of some individuals or groups yet creates friction with those of others. The immediate "winners" from changes in the Arctic are likely to include transport, construction, and oil and gas industries, as well as tourist operators and fishing enterprises, many of whom are based outside the region. As was noted by Duhaime (2004, 71), "While the industrial-scale natural resource exploitation creates considerable wealth . . . the resources generally belong to sources of capital outside the Arctic, which control the activities and profits." The other major beneficiaries of changes in the Arctic are consumers in general, who are likely to benefit from lower prices due to reduced transport costs for movement of goods and reduced energy costs.

While the "winners" from a warmer Arctic may be generally located outside the region, the immediate losers are likely to include many of the long-time residents of the region. Although some Arctic residents are benefiting from opportunities associated with resource extraction and development, many indigenous communities and groups whose livelihoods depend on existing ice and snow patterns and "cold conditions"—including associated flora and fauna—are likely to be harmed by changes in the Arctic (Adger 2004; ACIA 2004). Loss of sea ice is already threatening indigenous lifestyles and traditional hunting and herding cultures in communities throughout the Arctic (ACIA 2004; Ford et al. 2006). In Alaska and Siberia, communities are already facing problems associated with damage to buildings, pipelines, and infrastructure due to melting of permafrost (Goldman 2002; Lynas 2003).

Although there appears to be a clear pattern of winners and losers with respect to the Arctic, these patterns may be ephemeral under accelerating climate change. Over the long term, as climate change contributes to sea level rise, coastal areas and ports throughout the world will become increasingly vulnerable to shoreline erosion, flooding, and coastal storms (IPCC 2007a). Such hazards will be further magnified should rates of Arctic melting accelerate because of increased greenhouse gas emissions. The changing mass of the Greenland Ice Sheet has already raised concerns that current estimates of global sea level rise may be underestimated (Dowdeswell 2006). The consequences of extreme sea level rise would be overwhelming, since the world's coastal zones are already home to a significant part of humanity, and coastward migration is expected to increase throughout this century (Tol et al. 2006). Yet even a small rise in sea level will have dramatic consequences for shoreline erosion, creating complex challenges for coastal management

(Pilkey and Cooper 2004). Many of the groups and regions that will initially benefit from changes in the Arctic—shipping and transport industries, port areas, and urban consumers—may experience Beck's boomerang effect as large coastal cities and shipping infrastructure become subject to worsening storms and sea level rise. Those with economic interests in the north who are rejoicing at the possibility of an open Arctic and the trade opportunities that it creates may fail to recognize that sea level rise and changing weather patterns represent a dire threat to many of the consumers and trading partners that these countries envision.

Recognition of double exposure in the Arctic reveals critical challenges to human security. Yet the Arctic case also illustrates some of the opportunities and positive synergies that emerge via interactions between globalization and global environmental change. Global media coverage of the plight of the polar bear has raised public consciousness worldwide about the threat that climate change in the Arctic poses to species and ecosystems. Cross-border institutions such as the Arctic Council, a key partner in the Arctic Climate Impact Assessment (ACIA), provide a forum for intergovernmental cooperation, coordination, and knowledge exchange between Arctic states, indigenous communities, and other Arctic residents (Arctic Council 2006). Both global media and globalized institutions have also facilitated an attempt by the Inuit Circumpolar Conference (ICC) to hold large emitters accountable for damages resulting from climate change. In 2005 the ICC launched a formal petition with the Inter-American Commission on Human Rights, claiming that global warming caused by the United States is destroying Inuit culture and livelihoods (Inuit Circumpolar Conference 2005). Chapter 8 will further explore how the two processes may create new openings for enhancing human security.

Conclusion

Feedback double exposure emphasizes the linkages between global environmental change and globalization, drawing attention to how they may create feedbacks that amplify each process over time. These important feedback relationships are often overlooked in planning and development decisions, in part because they manifest over different time scales, but also because the full extent of linkages between the two processes is not recognized or acknowledged through the separate global change discourses. Some linkages may be neglected not out of ignorance or for lack of information but because valued outcomes differ, power relationships are uneven, and access to resources is controlled by various economic and political interests.

Feedback double exposure thus reveals a profound contradiction. Exploiting the benefits of global change is likely to accelerate processes of change, which will have widespread consequences that reverberate across space and time, with implications for long-term sustainability. The benefits of a warmer north may ultimately generate catastrophic losses as rising sea levels, increased climate variability, extreme climatic events, biodiversity loss, and ecosystem changes create new risks across the globe. As Beck puts it: "The basic insight lying behind all this is as simple as possible: everything which threatens life on this Earth also threatens the property and commercial interests of those who live *from* the commodification of life and its requisites" (Beck 1992, 38–39, italics in orginal).

Addressing feedback double exposure involves confronting unequal power relationships, challenging economic interests, and identifying values that jeopardize future sustainability. In particular, feedback double exposure draws attention to equity issues and the sustainability of current consumption practices. Responsibility for the loss of Arctic ice is not shared equally across human communities but can be disproportionately attributed to some nations, regions, and social groups. High consumers of energy, of raw materials, and of consumer goods share responsibility for ecosystem damage and social disruptions that result from growing international trade, climate change and sea level rise (Conca 2002; O'Brien and Leichenko 2006). In summary, feedback double exposure suggests a fundamental rethinking, not only of energy and material consumption practices but also of what we value and why (Princen et al. 2002; Leichenko and O'Brien 2006).

Human Security in an Era
of Global Change

8

Human security is achieved when and where individuals and
communities: have the options necessary to end, mitigate, or adapt
to threats to their human, environmental, and social rights; have
the capacity and freedom to exercise these options; and actively
participate in attaining these options.
—GECHS, *Science Plan: Global Environmental Changes and Human
Security Project*

With a consistent focus on human security, more integrated social
arrangements and more integrated global efforts can address the big
threats and make people more secure.
—Commission on Human Security, *Human Security Now*

Global environmental change and globalization are often viewed as separate and
distinct processes. Yet, as has been shown in this book, the two interact in numerous ways, creating linkages, feedbacks, and synergies across both space and time.
The tendency to view global change processes in isolation is reinforced by separate
and often competing discourses, each of which presents its own interpretation of
the processes as well as its own agenda for research and policy action. This separation of discourses not only limits the types of questions that are asked but also hides
some key opportunities for responding positively to change.

The double exposure framework provides a conceptual tool for investigating the interactions between global environmental change and globalization. The
framework demonstrates how the processes may together undermine human
security by exacerbating inequalities, increasing vulnerabilities, and accelerating
rates of change. Yet the framework also reveals how the linkages between the two

processes may be used to create synergies and new openings for enhancing human security. For example, growing global awareness about the threat of climate change is leading to many new initiatives and partnerships between NGOs, local communities, corporations, and multilateral agencies to promote alternative forms of energy.[1] Such initiatives, which may also reduce the costs of solar cells, wind power, and other technologies, can potentially bring energy to poorer countries at reasonable costs, possibly enhancing agricultural productivity and generating new sources of local income (Annan 2000). This chapter examines current efforts to address the challenges associated with each process, and it discusses the limitations of addressing each process in isolation. It then shows how positive synergies between the two processes can be used to create a more equitable, resilient, and sustainable future.

Addressing Global Change

The processes of global environmental change and globalization are generating a myriad of policies and actions by international institutions, national governments, NGOs, corporations, private foundations, religious organizations, civil society groups, and individuals. Many of these activities are implicitly or explicitly connected to the discourses and interpretations of global environmental change and globalization that were discussed in chapter 2, and most have been contested or criticized by proponents of alternative discourses. Here we will draw attention to the implications of double exposure for responses that address global change processes. Responses that address global change are grouped into three types, each corresponding to a key challenge that was revealed by one of the pathways of double exposure: (1) interventions intended to reduce outcome differentials; (2) interventions intended to reduce vulnerability; and (3) interventions intended to change the processes. We do not evaluate or prioritize specific interventions but instead identify contradictions that may emerge when each process is viewed in isolation. We emphasize the need for integrated responses that recognize the interactions between the two processes including the linkages between outcomes, contexts, and feedbacks.

Decreasing Outcome Differentials

Processes of global environmental change and globalization produce uneven outcomes across individuals, households, regions, and social groups. The pathway of outcome double exposure stresses that these uneven outcomes are interrelated through shared contextual environments that influence both exposure and the capacity to respond to each process. Yet most efforts to address uneven outcomes focus on the effects of a single type of global change, ignoring the overlaps and synergies with the other process.

Typical approaches to addressing the uneven outcomes of globalization include redistribution, adjustment assistance, and compensation for those who are negatively affected. Redistribution programs such as social welfare and unemployment assistance provide a minimum income, social and health services, and other types of support for unemployed workers. Adjustment assistance typically entails education, retraining, or relocation assistance. Compensation may include some type of direct financial remuneration, such as production subsidies that are intended to help farmers, firms, and other producers to be more competitive in the global economy. These approaches, all of which emphasize reduction of outcome differentials, have most commonly been initiated in response to the uneven effects of trade liberalization.

Uneven outcomes attributable to global environmental change are increasingly recognized, yet few specific measures have been taken to reduce outcome differentials. Instead, the consequences of shocks linked to global environmental change are typically dealt with through humanitarian aid and disaster relief. Responses to shocks such as floods or droughts include both material and logistical assistance that aims to save lives, reduce suffering, and help in recovery and reconstruction (Hilhorst 2004). The development and expansion of public and private insurance schemes, which are prevalent in advanced economies, have also been recommended as a mechanism to spread risks and reduce financial losses associated with global environmental change (Mills 2005; Linerooth-Bayer and Vári 2006).

The possibility of legal action and financial compensation to redress damage and loss of life is increasingly being discussed in relation to climate change (Tol and Verheyen 2004; Smith and Shearman 2006; Adger et al. 2006). The issue of compensation for loss of snow, glaciers, coral reefs, and other ecosystem features also forms part of a larger climate change equity movement (Athanasiou and Baer 2002; Adger 2004). Nonetheless, climate change compensation remains a highly contested issue (Adger et al. 2006). Adger (2004) elaborates on the difficulties of financial compensation in the Arctic, suggesting that climate change is a fundamentally unjust burden representing an externality from past and present polluters in other jurisdictional areas. He argues that a focus on security and danger as outcomes of climate change, as well as a clearer definition of the right to avoid such outcomes, would help to rectify these issues (Adger 2004). Farber (2007, 1631) considers the challenges involved in designing a fair and efficient system of compensation for climate change and suggests key lessons of the experiences from natural disaster compensation: "Private insurance may be inadequate to deal with large-scale impacts as opposed to more localized harms. Also, given the variety of institutional forms for providing compensation, we should not focus exclusively on the litigation system as a basis for compensation."

The compensation issue illustrates the political complexity of climate change responses, drawing attention to issues of justice and fairness between those who are responsible for most greenhouse gas emissions and those who are likely to experience the negative consequences (Adger et al. 2006; O'Brien and Leichenko 2006). Provision of funds to plan for climate change adaptation is consequently another preliminary measure to address the negative outcomes of climate change (Huq and Burton 2003). While such funds are under negotiation, the U.N. Global Environment Facility currently supports a number of programs to assist poor countries as they develop National Adaptation Programmes of Action (NAPAs) to identify their most urgent adaptation priorities (GEF 2006).

While interventions to reduce outcome differentials are important and necessary, they are likely to bring limited relief when they recognize only one process of global change, particularly in cases where another process is simultaneously exacerbating inequalities. These efforts are also likely to have limited success when underlying contextual environments remain unchanged. As has been emphasized throughout this book, the overlapping outcomes of the two processes are not random but in fact stem from common contextual factors. Moreover, the capacity to address uneven outcomes may be changing as a result of both processes. For example, the capacity of the public sector to address uneven outcomes is in many cases decreasing at the same time that there is an increasing demand for resources to address the negative outcomes. As Onis and Aysan (2000, 2) note, the globalization process itself "undermines the redistributive capacities of the nation-state which would otherwise even out the resulting income disparities [associated with globalization], at least to a certain extent." Strategies to decrease outcome differentials are necessary but insufficient responses to the challenges linked to double exposure.

Reducing Vulnerability

Many individuals, communities, regions, and groups are becoming more vulnerable to shocks and stressors as a result of both global environmental change and globalization. The pathway of context double exposure demonstrates how the processes are transforming biophysical, social, economic, political, institutional, technological, and cultural conditions, influencing both exposure and the capacity to respond to many types of change. Approaches to reducing vulnerability and improving resilience often focus on these contextual conditions, with a particular emphasis on increasing the capacity to adapt to changing conditions (Thompson et al. 2006; Kirby 2006).

Development aid has been considered by many as a means to both limit exposure and enhance the capacity to respond to global change processes (see Sachs 2005). This approach is well illustrated by the United Nations Millennium Development Goals (MDGs) for 2015. The goals include the eradication of poverty and hunger, universal primary education, gender equality and empowerment of women, a reduction of child mortality and improvements in maternal health, and combating HIV/AIDS and other diseases. The MDGs also emphasize environmental sustainability and the development of global partnerships for development. Transformations of the contextual environment on the scale envisioned by the MDGs are expected to reduce vulnerability of much of the world's population to many types of shocks and stresses. Yet the MDGs do not explicitly recognize how global environmental change and globalization interact to alter contextual conditions while at the same time influencing the frequency and magnitude of various types of extreme events (e.g., floods, economic shocks). The MDG emphasis on the least developed countries also fails to acknowledge new types of vulnerabilities within advanced economies, such as those revealed by the case studies of the Paris Heat Wave and Hurricane Katrina.

Other interventions to reduce vulnerability focus on changes in technological and institutional conditions. Technological solutions either may reduce exposure to shocks and stresses or may enhance response capacity. For example, technologies to reduce flood exposure can include the construction of sea walls, dams, and levees and the development of early warning systems. In the agricultural sector, technologies that reduce vulnerability to drought include new seed varieties and expansion of irrigation schemes. In relation to biodiversity loss, technological responses may include captive breeding programs and electronic monitoring systems. Institutional responses to biodiversity loss include bans on international trade through the Convention on International Trade in Endangered Species (CITES). Institutional approaches may also include adaptive management strategies and facilitation of collaborative learning and information sharing among public and private institutions (Lee 1999; Berkhout et al. 2006; Thompson 2006; Berkes 2007). Promotion of adaptation has received increased attention within discussions about climate variability and change, particularly since some change is expected to occur in the next decades regardless of mitigation efforts (Adger et al. 2007).

Transforming the contextual environment for individuals and communities can be considered an effective means of reducing vulnerability to global change processes. However, such approaches do not address the processes themselves, which are continually changing the context for both exposure and responses. Many initiatives to reduce vulnerability fail to recognize that global change processes are increasing the frequency and magnitude of extreme events, changing the nature of environmental shocks and their impacts on livelihoods (Schipper and Pelling 2006). Moreover, efforts to reduce vulnerability often tend to view globalization (e.g., opening of markets and expansion of international trade) as a positive step, without recognizing that these changes may also create new vulnerabilities.[2] Efforts that rely on private or technological solutions tend to be based on prior event histories, and they typically do not recognize new risks and uncertainties posed by processes such as climate change. The changing nature of risks poses new challenges to the insurance industry, which in itself may transform the context for new developments and investments (ABI 2004; Mills et al. 2005).

Changing the Processes

An accelerating pace of change is characteristic of both global environmental change and globalization. In many instances, changes associated with each process are occurring faster than communities or species can adapt, and at the same time they are undermining essential ecosystem services. The pathway of feedback double exposure illustrates how responses to the outcomes of global environmental change and globalization may create feedbacks that drive and accelerate global change processes. Efforts to manage the pace of change and promote more sustainable development trajectories often entail direct challenges to the processes themselves (IPCC 2007b). However, unless the linkages and feedbacks are considered, contradictory or negative consequences may result.

Both global environmental change and globalization processes are currently being challenged by multiple actors operating at various scales. Internationally, over 100 treaties and multinational agreements have been signed to protect the global environment. Some of these agreements address the driving factors behind environmental change. The Montreal Protocol, for example, limits the production of ozone-depleting chemicals such as chlorofluorocarbons (CFCs). Other international agreements, such as the Kyoto Protocol, represent a first step in controlling the rate and magnitude of climate change. The Biodiversity Convention, the Ramsar Convention on Wetlands, and the Convention on International Trade in Endangered Species (CITES) are other examples of international efforts to manage, slow, or halt environmental change.[3]

In addition to these multilateral efforts to manage environmental change, many social movements have emerged to challenge the driving forces behind the observed and projected changes (McDonald 2006). These movements may either support or criticize market environmentalism, which is increasingly pursued as a response to global environmental change. In the case of climate change, market environmentalism involves taking advantage of the differential costs of greenhouse gas reductions, both within specific countries and across national borders (Kruger and Pizer 2004; Bogdonoff and Rubin 2007). Carbon trading and offset programs utilize emerging regional and global markets for carbon, generating new opportunities for investors, banks, corporations, and entrepreneurs to profit through quota sales or carbon reduction schemes. Although projects associated with programs such as the Clean Development Mechanism are intended to improve energy access and enhance livelihoods in developing countries, many of these efforts have been criticized because of negative environmental and social consequences (Lohmann 2006). For example, the spread of monoculture plantations of eucalyptus trees, such as those cultivated in Minas Gerais, Brazil, to replace coal with charcoal in the production of pig iron, has desiccated swamps and streams, contaminated air and water, destroyed plant and animal species, and disrupted the livelihoods of small-scale farmers living nearby (Wysham 2005).

There are also many groups and actors around the world who are challenging globalization (Klein 2000; Kingsnorth 2003). As was discussed in chapter 2's review of globalization discourses, some of these efforts seek to halt the present course of globalization while others seek to make the course more fair and equitable. Examples of efforts to halt globalization range from antiglobalization protests, which have became a fixture at World Trade Organization (WTO) meetings in recent years, to the Local Currency Movement, which promotes locally based production and consumption (Helleiner 2002), to the World Social Forum, which provides a venue for groups and civil society movements concerned with development, social justice, equity, and the environment (Bello 2004).[4]

Strategies to transform rather than halt globalization direct attention to issues of fairness and equity, but particularly to the "unlevel playing field" that characterizes the differing contexts for promoting trade across the globe (Stiglitz 2006). Efforts to transform globalization are increasingly involving individual consumers in advanced economies. Such initiatives include fair trade labeling of food products, which identify foods produced under fair trade agreements, food miles campaign to raise awareness of the distance that many food products travel and the energy consumed, and geographic labeling of food sources to encourage consumption of local products.

While many of these efforts have succeeded in slowing or transforming processes of global change, they may also have unexpected negative consequences, or even contradictory effects, particularly when linkages or feedbacks to other processes are not taken into account. Responses intended to limit biodiversity loss, climate change, land use change, water scarcity, and other environmental issues may—via globalization processes—have unintended consequences for some individuals and communities. For example, the growing demand for renewable energy resources such as ethanol and other biofuels competes with food production and hence raises market prices for staple foods. Countries that increasingly depend on food imports as the result of trade liberalization may thus experience negative outcomes from such responses. Similarly, efforts to reduce "food miles" may limit export opportunities for farmers involved in "fair trade" production (Woodin and Lucas 2004).

The unintended consequences of various intervention efforts become apparent when we recognize the connections between the two global change processes. Yet addressing these unintended consequences raises broader ethical questions about all types of intervention strategies: Whose security is enhanced by these interventions, and whose security is compromised? Are uneven outcomes from *responses* to global change processes as significant as the uneven outcomes from the processes themselves? What types of values should serve as guidelines for addressing global change?[5] The double exposure framework can be used to identify interventions that are consistent with the goals of human security, including efforts "to protect the vital core of all human lives in ways that enhance human freedoms and human fulfilment" (Commission on Human Security 2003, 4). The framework shows that responses that address the context, outcomes, and processes together in an integrated manner are more likely to be consistent with these human security goals.

New Openings

We have emphasized throughout this book that global environmental change and globalization are transformative processes that are creating growing inequalities, increasing vulnerabilities, and accelerating the pace of change. Not surprisingly, human security remains an elusive goal for much of the world's population. Nevertheless, individuals and communities across the globe are increasingly claiming or reclaiming their rights to end, mitigate, or adapt to threats posed by double exposure (Kingsnorth 2003; Hardt and Negri 2004). Here we emphasize some of the opportunities associated with processes of global change. Although the double exposure framework is not intended to offer specific answers or solutions, recognition of positive interactions between global processes can reveal new openings that may promote human security.

Emerging coalitions that transcend traditional divides (North–South, race, class, and gender) and that challenge the inequities associated with climate change represent one example of how both processes can positively interact. Examples of these coalitions and new organizations include EcoEquity, which is devoted to the promotion of equitable and just solutions to climate change; the Climate Crisis Coalition, which includes representatives from environmental, labor, human rights, social justice, public health, indigenous rights, and religious groups in its efforts to stem climate change; and the Global Justice Ecology Project, which focuses on the connections between social justice and ecological awareness. Each of these coalitions recognizes the need for large-scale transformations of global energy infrastructure, coupled with dramatic improvements in human development to facilitate this transition (EcoEquity and Christian Aid 2006).

Globalization of media and communication technologies has also facilitated broad dissemination of information about global environmental issues. Although the globalization of "taste" for ocean fish in combination with better technologies for fishing are depleting stocks of nearly every species of large ocean fish, there is also growing awareness of this issue due to the role of the global media, which has led to the removal of shark and other species from menus in recent years. Increased environmental awareness is also contributing to global efforts to protect fisheries and fish stocks worldwide through certification programs and sustainable management practices (MSC 2006).

Opportunities to make positive connections between the two processes may also be illustrated in some of the emerging agricultural movements intended to alter consumption and production patterns. The fair trade movement aims to ensure that small farmers worldwide obtain fair prices for their harvests yet at the same time advocates environmental stewardship through promotion of organic and sustainable farming and land use methods (Woodin and Lucas 2004). The Slow Food Movement opposes homogenization of food tastes and the loss of cultural identities associated with food traditions while also supporting biodiversity, including protection of both domesticated and wild species (Petrini and Watson 2001). This movement also supports local farmers who are competing with large-scale industrial agriculture to maintain their livelihoods (Bové and Dufour 2001).

Recognition of growing connectivities associated with double exposure also suggests new possibilities for enhancing both social and ecological resilience in the face of emergencies and disasters. The globalization of communication technologies has played an important role in the emergence of new forms of humanitarian action (McDonald 2006). The spread of technologies such as early warning systems to remote areas of less-developed regions may increase preparedness and improve response times for disasters—a change that might have saved many lives during the tsunami that hit South and Southeast Asia in December 2004 (Mitchell 2006). Enhanced economic ties as the result of globalization of supply chains are also giving private sector actors, including multinational corporations, a greater stake in promoting resilience. Because these actors have vital production facilities in many countries, they are also playing a growing role in both disaster preparedness and disaster recovery (Bender 2006).

Political actions originating at the local level yet linked via transnational networks also offer opportunities to address collectively many global environmental issues. For example, hundreds of municipalities worldwide are implementing climate change mitigation strategies via participation in the Cities for Climate Protection campaign, which was initiated in 1991 by the International Council for Local Environmental Initiatives (ICLEI) (Bulkeley and Betsill 2003; Kousky and Schneider 2003; Slocum 2004; Young 2007). As of 2006, 261 cities in North America and more than 675 cities worldwide were participating in this campaign, setting greenhouse gases emission reduction targets and developing local action plans to reduce their emissions.

Communication of new ideas for adaptation to and mitigation of global environmental change is also increasingly facilitated via growing global networks (Moser and Dilling 2007; Young 2007). For example, Curitiba, Brazil, has become an icon for cities that are pursuing sustainable development and mitigation strategies for climate change (O'Meara Sheehan 2002). This city of 2.5 million residents has prioritized the development of public transit based on high-quality bus systems and a complementary package of measures that deemphasize cars, while also preserving green spaces and producing an array of economic, environmental, and social cobenefits (Wright and Fulton 2005). Curitiba's strategy has been emulated by many cities around the world and has in part influenced new Bus Rapid Transit (BRT) systems in Beijing, Jakarta, Seoul, and Leon (Mexico) as well as similar projects underway in Cape Town, Dar es Salaam, Hanoi, Lima, Mexico City, and Santiago (Wright and Fulton 2005).

While the above examples are modest in scope—and by no means unproblematic—they nonetheless reveal that new openings may exist for harnessing the interactions between global environmental change and globalization. Neither one is an inevitable process with predetermined outcomes. Although it is clear that societies will have to adapt to a changing climate, that many species are already threatened or endangered, and that many cultures and groups face challenges to their survival and security, there are still many windows of opportunity to "change the change." Recognition of double exposure highlights new openings for promoting outcomes that enhance human security.

Conclusion

Global environmental change and globalization are among the most transformative processes of the contemporary period. Both processes create uneven outcomes, and both vividly illustrate how actions taken in one place and time can influence outcomes in other places and at other times. Both processes also generate increasing uncertainties about the future. Many of these uncertainties are linked to the rate, magnitude, and type of changes that are occurring, but there are also uncertainties regarding capacities to respond to these changes.

Addressing the dynamic changes taking place in the world today requires moving beyond traditional intellectual boundaries to consider complex relationships between seemingly disparate issues and problems. The metaphor of double exposure emphasizes how two separate pictures may be overlapped— either unintentionally or intentionally—to form a new picture. This new picture is not static but instead is undergoing continual and rapid transformation as a result of interactions between global environmental change and globalization. Yet, as with photographic images, which are increasingly captured and manipulated in digital format, it is often difficult to "see" how resulting pictures are created via interacting processes of global change. The double exposure framework provides a conceptual tool for investigation of many types of interactions between these global processes. The three pathways highlighted in the framework draw attention to the spatial and temporal dimensions of risk and change, illustrating how the processes together pose significant new challenges for human security.

The framework also reveals how synergies between global environmental change and globalization create new openings for enhancing human security. It emphasizes that there is nothing inevitable about processes and outcomes of global change. Human decisions, values, and behaviors drive global change processes, and they shape the contexts in which these changes are experienced. Although we live in an era of profound change, the double exposure framework shows us that these changes can be used to create a more equitable, resilient, and sustainable future.

Notes

Chapter 1

1. There has been some debate as to whether localized problems such as industrial pollution, soil erosion, and changes in water quality and availability are indeed global environmental issues (see Buttel and Taylor 1994). Turner et al. (1991) suggest that local environmental changes are occurring extensively across the globe and thus represent important components of global environmental change.

2. Biodiversity incorporates ecosystem, species, and genetic diversity (Wilson 1988).

Chapter 2

1. A worldview can be considered a model that shapes perceptions. Some consider that there is only one true and correct worldview and that the challenge is to make it known and understood. Others consider there to be many legitimate worldviews which are continuously created and recreated as conditions change. The former position corresponds to the representation or Enlightenment paradigm, that is, that there is a single, empirical world or nature, while the latter corresponds to the postmodern or post-Enlightenment paradigm, that is, that the world and worldviews are not completely pre-given but are developed and constructed in historical contexts (Smith 1998; Wilber 2000).

2. Ironically, supporters of this argument often refer to a work of fiction by Michael Crichton (2004) to justify skepticism about climate change.

3. As Buttel and Taylor (1994) point out, this global discourse serves simultaneously as a scientific concept and a movement ideology.

4. The IPCC's Special Report on Economic Scenarios (Nakicenovic and Swart 2000) represents an attempt to develop consistent scenarios of future emissions trends, based on four narrative story lines that capture different development pathways: differing degrees of globalization versus regionalization, and differing emphasis on material wealth versus social and ecological values.

5. The Millennium Ecosystem Assessment (MA) was launched by the UN Secretary-General Kofi Annan in 2001, with major financial support from the UN, World Bank, and various national governments and private foundations.

6. The 1988 Montreal Protocol to regulate ozone-depleting chemicals is an example of an international regime that developed in response to the global threat of stratospheric ozone

depletion. In this case, science identified a problem that was global in nature, and the international community responded by agreeing to limit the production of chlorofluorocarbons and other ozone-depleting chemicals. The U.N. Framework Convention on Climate Change and the Biodiversity Convention represent attempts to replicate this success in order to address climate change and biodiversity loss. Interestingly, the shift toward addressing environmental risks at a global scale and the emergence of international environmental politics are often seen as important aspects of globalization (Beck 1992; Held et al. 1999; Speth 2003).

7. In terms of understanding vulnerability to environmental change, Brookfield (1999) argues that there is a need to consider the geophysical and human elements of the problem equally, without assuming that one or the other is dominant.

8. The coupled human-environment perspective emerged from the work of geographers such as Gilbert White, Robert Kates, Billie Lee Turner, and others (Kates and Burton 1986; Turner et al. 1990); the ecological perspective draws from the writings of Holling (1973) and Odum (1953) on ecology and systems theory.

9. The notion that rational actions will follow from better science is also featured in the early natural hazards in geography, from which much of the human-environment perspective on global change evolved (e.g., Burton et al. 1978).

10. The biophysical discourse, for example, has been criticized for applying positivist science to produce "technocratic" knowledge while ignoring the underlying socioeconomic processes that are transforming nature (Castree 2001). Likewise, the human-environment discourse has been criticized for ignoring questions of power and inequality, preferring to focus on narrow technological solutions over social change (Pelling 2001). While claiming to be integrative, it nonetheless holds society and nature as related yet distinct, conceiving of human-environment relationships in a dualistic, binary way (Castree 2001; Castree and MacMillan 2001). The language of "coupled" systems, for example, stresses interconnections but nonetheless implies that there are two separate systems that are coupled together.

11. As is noted by Merchant (2003), the critical view of scientific knowledge is not intended to suggest that the physical environment and environmental problems are not "real." Rather, the emphasis is on how knowledge and understanding of environmental change reflect different points of view and knowledge systems.

12. In labeling the discourse "benign" and "malignant," we are drawing from the discussion by Milanovic (2003) of the two faces of globalization.

13. Our categorization differs from those of others such as Held and McGrew (2002) in that it does not incorporate the "statist" or "protectionist" position, and we do not make fine differentiations between, for example, institutionalist and liberal internationalist positions but instead place these positions into the broader category of the benign discourse. We do not include statist and related positions, because these fall more squarely within the "skeptical" camp and suggest that national economies are predominant and that national governments remain the important power in the global economy (Held and McGrew 2002).

14. As Jacques (2006, 78) argues, "the environmental skeptical movement guards against paradigmatic changes to world dominant social values and institutions that guide the global accumulation and concentration of power."

15. In evaluating debates about North American economic integration, for example, Gilbert (2005) finds that the characterization of economic integration as inevitable and irreversible permeates policy discussions, commentaries by industrial representatives, and public testimony by prominent academics. She notes that "Policy makers, bankers, analysts, academics, and others contribute to the production of consent around certain kinds of economic 'truths' that work to structure events, practices, and interpretations" (Gilbert 2005, 209).

16. Beck (2003, 25) similarly argues that opponents tend to equate the dominant form of globalization with Americanization. His evidence for this argument is the fact that antiglobalization protests center primarily on protests against American military power, American market power, and American political and cultural influences.

17. As described by Sklair (2002), the members of this class typically share similar lifestyles, including patterns of higher education and consumption of luxury goods and services as well as use of exclusive clubs and resorts, private forms of travel and entertainment, and gated residences secured by armed surveillance. Many members of this class earn significant income from assets and investments in addition to their salaries and benefits. Members of the transnational capitalist class are most typically from advanced economies, but this class also includes large numbers from developing countries, including Brazil, China, and India.

18. Held's (2004, 16) proposal for a project of global social democracy emphasizes accountability, democracy and transparency in global governance, commitment to social justice and more equitable distribution of life chances, protection of local communities, regulation of the global economy, and engagement with stakeholders in corporate governance.

Chapter 3

1. At least two of the global environmental change discourses reviewed in Chapter 2 use the terminology of "stresses." Within the biophysical discourse, multiple stresses are defined to include pollution, population growth, and so forth. Within the human-environment discourse, multiple stresses include societal changes such urban growth and military conflict. Within both discourses, it now widely recognized that multiple, interacting stresses can push physical and ecological systems over thresholds to result in permanent changes to these systems (Steffen et al. 2004, 189; Turner et al. 2003a).

Chapter 4

1. The nationwide *Plan Canicule 2006* ("The Heat Wave Plan 2006") emphasizes anticipation and prevention as its main guideline. The fight against isolation, particularly of the elderly, is a main priority (Le Premier Ministre de la France 2006).

2. On the basis of research in Central Mexico, Klooster (2003) emphasizes that the environmental implications of economic development and globalization for forests are often place-specific and depend upon social institutions that coordinate rural people's environmental actions.

3. Sustainability, along with sustainable development, is a concept that has come under immense scrutiny since it was strongly embraced by governments, businesses, and NGOs in the late 1980s. Although it has been criticized for being vague, normative, and contradictory, the concept of sustainability has nevertheless changed the way that science is approached (Kates et al. 2001; John Robinson 2004).

4. Although it is common practice in urban literature to distinguish between cities in developing countries and those in advanced economies, these distinctions are becoming increasingly blurred as cities everywhere are becoming more homogeneous in physical form and in the lifestyles of their middle and upper-class residents (Leichenko and Solecki 2005).

Chapter 5

1. The EU, for example, has been shifting from subsidies linked to production in favor of subsidies that are linked to environmental, food safety, and animal welfare standards. These changes, which can be more broadly situated in the context of longer-term restructuring of

agricultural support (e.g., reform of the Common Agricultural Policy, or CAP, in Europe), have been characterized by some as hailing a new era of postproductivism in European agriculture (see Evans et al. 2002).

2. Many countries with advanced economies have emphasized nontrade concerns in ongoing WTO negotiations on agricultural trade liberalization, and the concept of "multifunctional agriculture" has been increasingly used as an argument in favor of protecting this sector. For example, Norway, Switzerland, the European Union, Korea, Japan, and some other countries have placed an emphasis on the national food security aspects of agriculture as well as on environmental benefits of viable rural areas.

3. In many developing countries, reductions in domestic agricultural supports were a critical part of structural adjustment programs initiated by the IMF during the 1980s and 1990s. These programs are frequently connected with neoliberalism.

4. In the case of NAFTA, for example, it was argued that farmers in the United States and Canada have a comparative advantage in capital-intensive grains while Mexican farmers have a comparative advantage in labor-intensive horticultural crops. American and Canadian producers were therefore expected to specialize in the production of capital-intensive corn and wheat while Mexican producers were expected to specialize in the production of labor-intensive fruit and vegetables.

5. Where uneven outcomes persist over time, blame is attributed to market failures.

6. For example, the WTO rules are widely perceived as favoring producers in advanced economies. Although direct subsidies to farmers—which are common in developing countries—are not allowed under the proposed WTO results, income payments to farmers are permitted (Miller 2004). This approach thus allows farmers in advanced economies to benefit from income payments but prevents farmers in many developing countries from obtaining subsidized inputs such as fertilizer.

7. Transaction costs, it is argued, impede equal participation in markets: "Powerful interests may wish to restructure institutions with the objective of serving their own short term interests, which may be achieved by increasing transactions costs through such devices as monopolies, taxes and other restrictions on contracting" (Kydd 2002, 6). Stevens (2003) notes that SPS standards are likely to have more of an effect on exports from Africa than would direct trade policies. Such standards are likely to hinder exports from Africa because producers will not be able to meet the required health standards. Changing standards entail more rigorous safety requirements in import markets (e.g., for pesticide residues), new health regulations (e.g., for bovine spongiform encephalopathy or BSE), and new forms of monitoring (e.g., keeping vaccination and movement records for each animal), all of which could put small farmers at a disadvantage (Stevens 2003).

8. As Kydd (2002, 2) notes, "For agriculture, horticulture and floriculture, the activities in which smallholders can engage, the end markets are dominated by large retaining firms, which compete among themselves on continuing minor innovations in products and packaging, on maintaining strict quality criteria and on price."

9. Some researchers find evidence that the poor may, overall, gain from liberalization but that substantial and disparate groups often lose and therefore need compensation (IFAD 2001). Equality—including equal access to markets and to asset control—greatly helps the poor to benefit from liberalization. Rural inequalities, in contrast, tend to result in higher food prices' penalizing many of the rural poor who are net food buyers. "People with little education, few roads or contacts, or not speaking a majority language, are especially likely to be 'stuck' as immobile losers" (IFAD 2001, 10).

10. The case studies in the two districts, carried out in 2002–2003, employed a variety of participatory rural appraisal techniques that allowed us to cross-check responses to key research questions across multiple sources of data. Our team interviewed government

officials in the district administration to determine which agriculturally relevant state policy reforms had been implemented since the liberalization process began in 1991. We then
selected villages for local-level studies on the basis of secondary statistics of socioeconomic
and climatic conditions in various parts of the district, as well as on discussions with local
experts from governmental and nongovernmental organizations. Finally, household surveys were carried out to assess how agricultural reforms have influenced farmers' and agricultural laborers' livelihoods and ability to cope with calamities such as drought (O'Brien
et al. 2004).

11. Anantapur has repeatedly made world headlines since the late 1990s because of its
high rate of farmer suicides. The suicides are linked to repeated multi-year droughts combined with fluctuating oilseed prices, which have contributed to high levels of indebtedness
among small farmers.

12. Within India, lower castes and scheduled tribes (tribal population groups living
in remote areas) represent some of the poorest groups, and their semisubsistence economies are likely to be adversely affected by climate variability and change. As Shurmer-Smith
(2000) notes, poverty amounts to malnutrition in India, and India's poor are too poor to
survive even short-term hardship.

13. "The argument can be constructed, although it is difficult to substantiate in practice,
that it is rural women who carry a disproportionate share of the negative consequences of
reform, that is, the welfare of rural women declines relative to that of men." (Ellis 2000, 171).

14. Agarwal (1998, xvi) summarizes the factors that constrain women from exercis-
ing legal claims over land: "patrilocal post-marital residence and village exogamy, strong
opposition from male kin, the social construction of gender needs and roles, low levels of
female education, and male bias and dominance in administrative, judicial, and other public
decision-making bodies at all levels."

Chapter 6

1. Many of the most memorable geophysical disasters historically—from the destruction of Pompeii by a volcanic eruption in 79 AD to the San Francisco, Mexico City, and
Istanbul earthquakes in the twentieth century—have occurred in cities.

2. Many ideas from the early hazards literature (e.g., Burton et al. 1978) have been
applied to understanding vulnerability to global environmental change, with particular
emphasis on the causes of vulnerability, whether biophysical, social, or institutional.

3. Referring to the financial crisis in Indonesia, Frankenberg et al. (2002, 4) emphasize
the uneven outcomes: "For some, the impacts may have been devastating, but for others the
crisis has likely brought new opportunities. Exporters, export producers, and food producers fared far better than those engaged in the production of services and non-tradeables or
those on fixed incomes."

4. Kirby (2006, 33–35) identifies five main sources of risk associated with globalization: (1) the interconnected nature of the system, so that crises in one part reverberate almost
instantaneously throughout the global system; (2) new financial instruments such as deriva-
tives, that have been developed to anticipate and profit from price movements in currencies,
commodities, and equities; (3) lack of regulation of international financial flows; (4) the fact
that impacts are not limited to those who consciously gamble on the system; and (5) the
system as a whole being driven by a complex mixture of rational calculation and irrational
activity that heightens its unpredictability.

5. A worst-case scenario, based on the melting of the West-Antarctic Ice Sheet, involves
an increase in sea levels of up to five meters by 2130 (Tol et al. 2006).

6. Within advanced economies such as the United States or the countries of the EU, state responsibilities typically extend to social welfare programs, unemployment insurance, flood and hazard insurance, and crop insurance.

7. Van Vliet (2002) points out that because many new urban environmental initiatives center around issues such as land use, watershed protection, waste reduction, and recycling, all of which originate at the local rather than national level, devolution may sometimes open new possibilities for action to protect local environments.

8. In some cases the two processes together can lead to contradictory responses. Globalization, for example, often demands both specialization and continuous innovation in order to maintain competitiveness (Friedman 2005). Yet specialization often limits livelihood diversity at the same time that such diversity is increasingly needed to respond to different kinds of shocks and stresses.

9. The U.S. Congress is evaluating the decommission and closure of MRGO, because of environmental damage, high maintenance costs, limited economic benefits, and its possible role in contributing to hurricane vulnerability (Carter and Stern 2006).

10. Glasmeier's metric of persistent poverty includes various comparative measures of economic health and well-being at the county level (Glasmeier 2005). Cutter et al.'s (2006) social vulnerability index is based on a wide range of socioeconomic and demographic variables and includes metrics of wealth, race, age, economic structure, gender, and other factors thought to influence vulnerability to environmental hazards (Cutter et al. 2003).

11. Just prior to Katrina, sales tax revenue is reported to have accounted for close to one third of the city's tax base (Rivlin 2005).

12. The largest employment sector in New Orleans in 2005 was also service related and included social services, health care, and education, which accounted for more than 27 percent of the city's jobs. For the U.S. overall, services, health care, and education accounted for approximately 20 percent of jobs in 2005 (U.S. Census Bureau 2005).

13. Changes at FEMA since 2001 have included budget reductions, loss of cabinet status, and replacement of experienced staffers with political appointees. As Rodrigue (2006) describes it, "As FEMA entered the 2005 hurricane season, it was staffed at the highest levels by inexperienced managers, had seen the frustrations of the core staffers translate into resignations and the coring out of institutional memory, the erosion of the thick ties among FEMA staffers and their counterparts in state and local agencies, and its re-integration into at least two disconnected chains of command." These factors led to a situation in which core functions were undermined and hazard management turned into disaster.

14. Among the dead were 154 hospital and nursing home patients, the majority of whom died while awaiting rescue. "By the scores, people without choice of whether to leave or stay perished in New Orleans, trapped in health care facilities and in many cases abandoned by their would-be government rescuers" (Rohde et al. 2005).

Chapter 7

1. Globalization has been defined by Held and McGrew (2000, 2) as "a widening, deepening and speeding up of worldwide interconnectedness in all aspects of contemporary social life, from the cultural to the criminal, the financial to the spiritual."

2. The emphasis on the local environment is significant, as increased incomes and higher levels of development imply high per capita consumption and hence the potential for a greater net impact on the global environment (McGranahan et al. 2001; Marcotullio and Lee 2003; McGranahan 2007).

3. In addition to the NSR, another large shipping route is being developed, called the Northern Maritime Corridor (NMC). It aims to facilitate the transport of goods within the North Sea region to connect the North Sea basin with the Northern Periphery area (see note 9 to this chapter).

4. According to Østreng (1999), the NSR can be likened to the Northeast Passage that stretches from London to Japan, where "it is seldom necessary to negotiate more than 400 nautical miles of difficult ice conditions, i.e., 5–6% of the total freight distance. . . . This fact has given birth to vivid visions and perspectives, promising geopolitical and geoeconomic benefits with far-reaching consequences" (Østreng 1999, 10).

5. Østreng (1999, 7) points out that "[a] sea route, in the functional tradition, is a trading link—actual or potential—between towns and cities with harbours, with loading, service and reception facilities, transport networks, sizeable populations etc. Neither the Bering Strait nor Novaya Zamlya meets any of these criteria. . . . On this backdrop, it has been argued that the NSR should be defined functionally as connecting towns and cities of the Pacific side of the Russian Far east with those in the European part of Russia, for instance Murmansk." With Vladivostok as the Russian eastern end point, the neighboring countries of Japan, North Korea, South Korea, and China can be considered functional end points of the NSR (Østreng 1999, 9). On the western side, the functional definition of the NSR extends it to the coast of northern Norway.

6. Under current environmental and technological conditions, navigation through the NSR is considered a profitable alternative for some types of transport. However, to reach its full potential as a trade route, some major technological, financial, and political problems will need to be solved (Granberg 1998; INSROP 1998). Some of the challenges are linked to Russian tariff policies, investments to develop and maintain infrastructure, the jurisdictional status of the Russian Arctic straits, and the costs of building a fleet of operationally and environmentally sound state-of-the-art ice-classified vessels (INSROP 1998).

7. Historically, "no NSR region is completely ice-free during the short summer (although the southwestern Kara Sea is very close at 95% ice-free in September). This fact alone would seemingly prohibit non-ice strengthened ships from making passage along the NSR" (Brigham et al. 1999, 69). The Siberian seas experienced significant regional reductions in sea ice during the 1990s (Brigham et al. 1999). This recent decrease in sea ice may lengthen access to the eastern NSR from the Bering Strait to 150 days, that is, into November and sometimes as late as December.

8. The northern boundary of the NSR falls largely within the Russian 200-mile economic zone in the Arctic Ocean (Østreng 1999). The Russian Federation faces some natural impediments to transforming the NSR into a viable commercial route. These include climatic conditions, ice, shoals, and polar darkness, as well as permanent features such as the shallowness of the waters of the continental shelf.

9. The Northern Maritime Corridor (NMC) is the term for a sea-based transportation corridor that stretches from northern Norway and northwest Russia to the Continent, connecting the coastal areas of the North Sea and the Northern Periphery. The NMC project, which ran from 2002 to 2008, included participation of 20 regions in eight countries (Germany, the Netherlands, Belgium, Scotland, Denmark, Iceland, Sweden, and Norway). The vision of the NMC is to develop a means of efficient, safe, and sustainable transportation that connects coastal areas and enhances regional development in the North Sea region and the Northern Periphery area. This vision is pursued through the establishment of regional maritime clusters, comprising industries, ship owners, transporters, and port authorities, creating an arena where these can meet and develop new services in the corridor (Northern Maritime Corridor 2007).

Chapter 8

1. A number of new climate change partnerships and initiatives were announced in 2007, some with significant levels of financial support. Examples include a U.S. $100 million partnership sponsored by HSBC bank, in collaboration with The Climate Group, Earthwatch Institute, Smithsonian Tropical Research Institute (STRI), and the World Wildlife Fund (Osborne 2007), and the formation of an Urban Climate Change Research Network with participants from universities, local governments, multilateral agencies, NGOs, and the private sector (Earth Institute News 2007).

2. The last of the Millennium Development goals—to develop a global partnership for development—is aimed, for example, at supporting and directing the process of globalization in that it makes a call to "Develop further an open trading and financial system that is rule-based, predictable and non-discriminatory." (UN 2005b).

3. An important limitation of many of these multilateral treaties is that they limit participation to sovereign states and exclude other key actors such as NGOs, local governments, members of civil society, and the private sector.

4. The World Social Forum, held annually in different cities throughout the world, is described as "an open meeting place for reflective thinking, democratic debate of ideas, formulation of proposals, free exchange of experiences and inter-linking for effective action, by groups and movements of civil society that are opposed to neo-liberalism and to domination of the world by capital and any form of imperialism, and are committed to building a society centred on the human person." (World Social Forum India 2007).

5. Interventions may be evaluated on the basis of principles of utilitarianism, which focus on achieving the greatest amount of good for the greatest number of people, versus Rawls' (1972) notions of egalitarianism, which suggests that those who are negatively affected by various interventions must also be compensated (O'Brien and Leichenko 2003).

References

Aall, Carlo, and Ingrid Thorsen Norland. 2002. *The Ecological Footprint of the City of Oslo— Results and Proposals for the Use of Ecological Footprint in Local Environmental Policy.* Oslo: Program for Research and Documentation for a Sustainable Society (ProSus) and Western Norway Research Institute.

ABI. 2004. *A Changing Climate for Insurance: A Summary Report for Chief Executives and Policymakers.* London: Association of British Insurers.

Abrahamson, Mark. 2004. *Global Cities.* New York: Oxford University Press.

ACIA (Arctic Climate Impact Assessment). 2004. *Impacts of a Warming Arctic: Arctic Climate Impact Assessment.* Cambridge: Cambridge University Press.

Adger, W. Neil. 1999. Social Vulnerability to Climate Change and Extremes in Coastal Vietnam. *World Development* 27(2): 249–269.

———. 2000a. Institutional Adaptation to Environmental Risk under Transition in Vietnam. *Annals of the Association of American Geographers* 90(4): 738–758.

———. 2000b. Social and Ecological Resilience: Are They Related? *Progress in Human Geography* 24: 347–364.

———. 2004. Commentary: The Right to Keep Cold. *Environment and Planning A* 36(10): 1711–1715.

Adger, W. Neil, and Mick Kelly. 1999. Social Vulnerability to Climate Change and the Architecture of Entitlements. *Mitigation and Adaptation Strategies for Global Change* 4(3–4): 253–266.

Adger, W. Neil, Tor A. Benjaminsen, Katrina Brown, and Hanne Svarstad. 2001. Advancing a Political Ecology of Global Environmental Discourses. *Development and Change* 32(4): 681–715.

Adger, W. Neil, Saleemul Huq, Katrina Brown, Declan Conway, and Mike Hulme. 2003. Adaptation to Climate Change in the Developing World. *Progress in Development Studies* 3(3): 179–195.

Adger, W. Neil, and Nick Brooks. 2003. Does Global Environmental Change Cause Vulnerability to Disaster? In *Natural Disasters and Development in a Globalizing World,* ed. Mark Pelling, 19–42. London: Routledge.

Adger, W. Neil, Terry P. Hughes, Carl Folke, Stephen R. Carpenter, and Johan Rockström. 2005. Social-Ecological Resilience to Coastal Disasters. *Science* 309: 1036–1039.

Adger, W. Neil, Jouni Paavola, Saleem Huq, and M. J. Mace, eds. 2006. *Fairness in Adaptation to Climate Change*. Cambridge, MA: MIT Press.

Adger, W. Neil, Shardul Agrawala, Monirul Mirza, Cecilia Conde, Karen O'Brien, Juan Puhlin, Roger Pulwarty, Barry Smit, and Kyoshi Takahashi. 2007. Assessment of Adaptation Practices, Options, Constraints and Capacity. In *IPCC Fourth Assessment Report: Impacts, Vulnerability, and Adaptation*, ed. M. L. Parry, O. F. Canziani, J. P. Palutikof, P. J. van der Linden, and C. E. Hanson, 717–743. Cambridge: Cambridge University Press.

Adhikari, Ratnakar. 2000. Agreement on Agriculture and Food Security: South Asian Perspective. *South Asia Economic Journal* 1(2): 43–61.

Agarwal, Bina. 1998. *A Field of One's Own: Gender and Land Rights in South Asia*. Cambridge: Cambridge University Press.

Aggarwal, Pramod K., and R. K. Mall. 2002. Climate Change and Rice Yields in Diverse Agro-Environments of India. II. Effect of Uncertainties in Scenarios and Crop Models on Impact Assessment. *Climatic Change* 52(3): 331–343.

Aggarwal, Rimjhim M. 2006. Globalization, Local Ecosystems, and the Rural Poor. *World Development* 34: 1405–1418.

Agyeman, Julian, Robert D. Bullard, and Bob Evans, eds. 2003. *Just Sustainabilities: Development in an Unequal World*. Cambridge, MA: MIT Press.

AHDR. 2004. *Arctic Human Development Report*. Akureyri: Steffanson Arctic Institute.

Aide, T. Mitchell, and H. Ricardo Grau. 2004. Globalization, Migration, and Latin American Ecosystems. *Science* 305: 1915–1916.

Alley, Richard B., Peter U. Clark, Philippe Huybrechts, and Ian Joughlin. 2005. Ice-Sheet and Sea-Level Changes. *Science* 310: 456–460.

AMAP. 2005. *AMAP Assessment 2002: Heavy Metals in the Arctic*. Arctic Monitoring and Assessment Programme (AMAP). Oslo.

Amin, Samir. 1997. *Capitalism in the Age of Globalization*. London: Zed Books.

Anderson, Kym, and Will Martin. 2005. Agricultural Trade Reform and the Doha Development Agenda. *The World Economy* 29(9): 1301–1327.

Anderson, Sarah, ed. 2000. *Views from the South: The Effects of Globalization and the WTO on Third World Countries*. London: Food First Books, The International Forum on Globalization.

Anisimov, O., and B. Fitzharris. 2001. Polar Regions (Arctic and Antarctic). In *Climate Change 2001: Impacts, Adaptation, and Vulnerability*. Contribution of Working Group II to the Third Assessment Report of the Intergovernmental Panel on Climate Change, ed. J. J. McCarthy, O. F. Canziani, N. A. Leary, D. J. Dokken, and K. S. White, 801–841. Cambridge: Cambridge University Press.

Annan, Kofi. 2000. Sustaining the Earth in the New Millennium. *Environment* 42(8): 20–30.

Araújo, Miguel B., and Carsten Rahbek. 2006. How Does Climate Change Affect Biodiversity? *Science* 313: 1396–1397.

Arctic Council. 2006. About the Arctic Council. Oslo: Arctic Council Secretariat. Accessed May 29, 2007 <http://www.arctic-council.org/>.

Arctic Energy Summit. 2006. *Extractive Theme*. Accessed March 8, 2007 <https://www.confmanager.com/main.cfm?cid=680&nid=5792>.

Arvai, Joseph, Gavin Bridge, Nives Dolsak, Robert Franzese, Tomas Koontz, April Luginbuhl, Paul Robbins, Kenneth Richards, Katrina Smigh Korfmacher, Brent Sohngen, James Tansey, and Alexander Thompson. 2006. Adaptive Management of the Global Climate Problem: Bridging the Gap between Climate Research and Climate Policy. *Climatic Change* 78: 217–225.

Athanasiou, Tom, and Paul Baer. 2002. *Dead Heat: Global Justice and Global Warming.* New York: Seven Stories Press.

Atkins, Peter, and Ian Bowler. 2001. *Food in Society: Economy, Culture, Geography.* London: Arnold Publishers.

Baddeley, Michelle, Kirsty McNay, and Robert Cassen. 2006. Divergence in India: Income Differentials at the State Level: 1970–1997. *Journal of Development Studies* 42(6): 1000–1022.

Bankoff, Greg, Georg Frerks, and Dorothea Hilhorst, eds. 2004. *Mapping Vulnerability: Disasters, Development & People.* London: Earthscan.

Barlow, Maude, and Tony Clarke. 2002. *Blue Gold. The Fight to Stop the Corporate Theft of the World's Water.* New York: The New Press.

Barnett, Anthony, David Held, and Caspar Henderson. 2007. *Debating Globalization.* Cambridge, UK: Polity Press in association with openDemocracy.

Barnett, Jon. 2001a. Adapting to Climate Change in Pacific Island Countries: The Problem of Uncertainty. *World Development* 29(6): 977–993.

———. 2001b. *The Meaning of Environmental Security: Ecological Politics and Policy in the New Security Era.* London: Zed Books.

Barnett, Jon, and Jonathan Pauling. 2005. The Environmental Effects of New Zealand's Free-Market Reforms. *Environment, Development and Sustainability* 7: 271–289.

Beall, Jo. 2002. Living in the Present, Investing in the Future—Household Security Among the Poor. In *Urban Livelihoods: A People-Centred Approach to Reducing Poverty,* ed. Carole Rakodi and Tony Lloyd-Jones, 71–87. London: Earthscan.

Bebbington, Anthony. 2001. Globalized Andes? Livelihoods, Landscapes and Development. *Ecumene* 8(4): 414–436.

Beck, Ulrich. 1992. *Risk Society: Towards a New Modernity.* London: Sage Publications.

———. 1999. *World Risk Society.* Cambridge, UK: Polity Press.

———. 2000. *What Is Globalization?* Cambridge, UK: Polity Press.

———. 2003. Rooted Cosmopolitanism: Emerging from a Rivalry of Distinctions. In *Global America: The Cultural Consequences of Globalization,* ed. Ulrich Beck, Nathan Sznaider, and Rainer Winter, 15–29. Liverpool: Liverpool University Press.

Belliveau, Suzanne, Barry Smit, and Ben Bradshaw. 2006. Multiple Exposures and Dynamic Vulnerability: Evidence from the Grape Industry in the Okanagan Valley, Canada. *Global Environmental Change* 16(4): 364–378.

Bello, Walden. 2004. *Deglobalization: Ideas for a New World Economy.* London: Zed Books.

Bender, Steven. 2006. Dependency and Interdependency in Disasters and Globalization. Paper presented at Third Annual MaGrann Conference, "The Future of Disasters in a Globalizing World." New Brunswick, NJ: Rutgers University, April 21–22, 2006.

Beniston, Martin. 2004. The 2003 Heat Wave in Europe: A Shape of Things to Come? An Analysis Based on Swiss Climatological Data and Model Simulations. *Geophysical Research Letters* 31: LO2202.

Berkes, Fikret. 2007. Understanding Uncertainty and Reducing Vulnerability: Lessons from Resilience Thinking. *Natural Hazards* 41: 283–295.

Berkes, F., T. P. Hughes, R. S. Steneck, J. A. Wilson, D. R. Bellwood, B. Crona, C. Folke, L. H. Gunderson, H. M. Leslie, J. Norberg, M. Nyström, P. Olsson, H. Österblom, M. Scheffer, and B. Worm. 2006. Globalization, Roving Bandits and Marine Resources. *Science* 311: 1557–1558.

Berkhout, Frans, Julia Hertin, and David M. Gann. 2006. Learning to Adapt: Organisational Adaptation to Climate Change Impacts. *Climatic Change* 78: 135–156.

Berkhout, Frans, Melissa Leach, and Ian Scoones, eds. 2003. *Negotiating Environmental Change: New Perspectives from Social Science*. Cheltenham: Edward Elgar.

Bernard, Andrew, and J. Bradford Jenson. 2000. Understanding Increasing and Decreasing Wage Inequality. In *The Impact of International Trade on Wages*, ed. Robert Feenstra, 227–267. Chicago: National Bureau of Economic Research and University of Chicago Press.

Bhalla, G. S., ed. 1994. *Economic Liberalization and Indian Agriculture*. New Delhi: Institute for Studies in Industrial Development.

Blaikie, Piers, and Harold Brookfield. 1987. *Land Degradation and Society*. London: Methuen.

Blinder, Alan S. 2006. Offshoring: The Next Industrial Revolution. *Foreign Affairs* 85(2): 113–128.

Bogdonoff, Sondra, and Jonathan Rubin. 2007. The Regional Greenhouse Gas Initiative: Taking Action in Maine. *Environment* 49(2): 9–16.

Bohle, H. G., T. E. Downing, and M. J. Watts. 1994. Climate Change and Social Vulnerability. *Global Environmental Change* 4(1): 37–48.

Bohman, James. 2007. *Democracy across Borders: From Dêmos to Dêmoi*. Cambridge, MA: MIT Press.

Boulding, Kenneth E. 1978. *Stable Peace*. Austin: University of Texas Press.

Bouwman, A. F., T. Kram, and K. Klein Goldewijk, eds. 2006. *Integrated Modelling of Global Environmental Change: An Overview of IMAGE 2.4*. Bilthoven, The Netherlands: Netherlands Environmental Assessment Agency (MNP). Accessed June 15, 2007 <http://www.mnp.nl/bibliotheek/rapporten/500110002.pdf>.

Bové, José, and François Dufour. 2001. *The World Is Not for Sale: Farmers Against Junk Food*. London: Verso.

Boyer, Jeff, and Aaron Pell. 1999. Mitch in Honduras: A Disaster Waiting to Happen. *NACLA Report on the Americas*: 36–43.

Bredahl, Maury E., Nicole Ballenger, John C. Dunmore, and Terry L. Roe, eds. 1996. *Agriculture, Trade, and the Environment: Discovering and Measuring the Critical Linkages*. Boulder, CO: Westview Press.

Brenner, Neil, and Nik Theodore. 2002. Cities and the Geographies of "Actually Existing Neoliberalism." *Antipode* 34(3): 349–379.

Brewer, Thomas L. 2003. The Trade Regime and the Climate Regime: Institutional Evolution and Adaptation. *Climate Policy* 3(4):329–341.

Bridge, Gavin. 2002. Grounding Globalization: The Prospects and Perils of Linking Economic Processes of Globalization to Environmental Outcomes. *Economic Geography* 78(3): 361–386.

Brigham, Lawson W., Vladimir D. Grishchencko, and Kazuhiko Kamisaki. 1999. The Natural Environment, Ice Navigation and Ship Technology. In *The Natural and Societal Challenges of the Northern Sea Route. A Reference Work*, ed. Willy Østreng, 47–120. Dordrecht: Kluwer Academic Publishers.

Brinkley, Douglass. 2006. *The Great Deluge: Hurricane Katrina, New Orleans, and the Mississippi Gulf Coast*. New York: HarperCollins.

Brookfield, Harold. 1999. Environmental Damage: Distinguishing Human from Geophysical Causes. *Environmental Hazards* 1(1): 3–11.

Bruinsma, Jelle, ed. 2003. *World Agriculture: Towards 2105/2030—An FAO Perspective*. London: Earthscan.

Bulkeley, Harriet, and Michele M. Betsill. 2003. *Cities and Climate Change: Urban Sustainability and Global Environmental Governance*. London: Routledge.

Burton, Ian, Robert W. Kates, and Gilbert F. White. 1978. *The Environment as Hazard*. New York: Oxford University Press.

Buttel, Frederick, and Peter Taylor. 1994. Environmental Sociology and Global Environmental Change: A Critical Assessment. In *Social Theory and the Global Environment*, ed. Michael Redclift and Ted Benton, 228–255. London: Routledge.

Cannon, Terry. 1994. Vulnerability Analysis and the Explanation of "Natural" Disasters. In *Disasters, Development, and Environment*, ed. Ann Varey, 13–30. Chichester: John Wiley.

Carter, Nicole T., and Charles V. Stern. 2006. *Mississippi River Gulf Outlet: Issues for Congress*. Congressional Research Services. U.S. Library of Congress (August 6, 2006). Accessed January 10, 2007 <http://ncseonline.org/NLE/CRSreports/06Sep/RL33597.pdf>.

Carter, T. R., M. L. Parry, H. Harasawa, and S. Nishioka. 1994. *IPCC Technical Guidelines Assessing Climate Change Impacts and Adaptations*. University College London, Centre for Global Environmental Research, Japan.

Cash, David W., William C. Clark, Frank Alcock, Nancy M. Dickson, Noelle Eckley, David H. Guston, Jill Jäger, and Ronald B. Mitchell. 2003. Knowledge Systems for Sustainable Development. *Proceedings of the National Academy of Sciences (PNAS)* 100: 8086–8091.

Castells, Manuel. 1998. *End of Millennium*. Malden, MA: Blackwell.

Castree, Noel. 1995. The Nature of Produced Nature: Materiality and Knowledge Construction in Marxism. *Antipode* 27: 12–48.

———. 2001. Socializing Nature: Theory, Practice, and Politics. In *Social Nature: Theory, Practice, and Politics*, ed. Noel Castree and Bruce Braun, 1–21. Malden, MA: Blackwell Publishers.

Castree, Noel, and Tom MacMillan. 2001. Dissolving Dualisms: Actor Networks and the Reimagination of Nature. In *Social Nature: Theory, Practice, and Politics*, ed. Noel Castree and Bruce Braun, 208–224. Malden, MA: Blackwell Publishers.

Center for Social Inclusion. 2006. *Race to Rebuild: The Color of Opportunity and the Future of New Orleans*. New York: Center for Social Inclusion. Accessed June 15, 2007 <http://www.centerforsocialinclusion.org/PDF/racetorebuild.pdf>.

Chand, Ramesh, P. 2004. *Impact of Trade Liberalization and Related Reforms on India's Agricultural Sector, Rural Food Security, Income and Poverty*. Global Development Network Library: Research Papers from GDN Activities, Annual Conference. Accessed February 20, 2007 <http://www.gdnet.org/middle.php?oid=804&zone=&action=doc&doc=10280>.

Chand, Ramesh, Dayanatha Jha, and Surabhi Mittal. 2004. WTO and Oilseeds Sector: Challenges of Trade Liberalization. *Economic and Political Weekly* 39 (Feb 7, 2004): 533–537.

Chapin, F. S., G. Peterson, F. Berkes, T. V. Callaghan, P. Angelstam, M. Apps, C. Beier, Y. Bergeron, A.–S. Crépin, K. Dannell, T. Elmqvist, C. Folke, B. Forbes, N. Frescho, G. Juday, J. Niemelä, A. Shvidenko, and G. Whiteman. 2004. Resilience and Vulnerability of Northern Regions to Social and Environmental Change. *Ambio* 33: 344–349.

Chase, Thomas N., Klaus Wolter, Roger A. Pielke, Sr., and Ichtiaque Rasool. 2006. Was the 2003 European Summer Heat Wave Unusual in a Global Context? *Geophysical Research Letters*. 33(23): L23709, doi:10.1029/2006GL027470.

Chaudhury, P. 1998. Food Security and Globalisation of Indian Agriculture. In *Economic Liberalisation in India*, ed. Biswaijt Chatterjee, 256–274. Calcutta: Allied Publishers Limited.

Christensen, Tom, and Per Lægreid, eds. 2001. *New Public Management: the Transformation of Ideas and Practice*. Hampshire, UK: Ashgate.

Church, J. A., J. M. Gregory, P. Huybrechts, M. Kuhn, K. Lambeck, M. T. Nhuan, D. Qin, and P. L. Woodworth. 2001. Changes in Sea Level. In *Climate Change 2001: The Scientific Basis. Contribution of Working Group 1 to the Third Assessment Report of the Intergovernmental Panel on Climate Change*, ed. J. T. Houghton et al., 639–694. Cambridge: Cambridge University Press.

Cieslewicz, David J. 2002. The Environmental Impacts of Sprawl. In *Urban Sprawl: Causes, Consequences and Policy Response*, ed. Gregory D. Squires, 23–38. Washington, DC: The Urban Institute.

Clapp, Jennifer. 2006. WTO Agriculture Negotiations: Implications for the Global South. *Third World Quarterly* 27(4): 563–577.

Clark, William C., Paul J. Crutzen, and Hans J. Schellnhuber. 2005. *Science for Global Sustainability: Toward a New Paradigm*. CID Working Paper No. 120. Cambridge, MA: Science, Environment and Development Group, Center for International Development, Harvard University.

Coast 2050. 1998. *Coast 2050: Toward a Sustainable Coastal Louisiana*. Louisiana Coastal Wetlands Conservation and Restoration Task Force and Wetlands Conservation and Restoration Authority. Louisiana Department of Natural Resources. Baton Rouge.

Cole, Alistair, and Glyn Jones. 2005. Reshaping the State: Administrative Reform and New Public Management in France. *Governance: An International Journal of Policy, Administration, and Institutions* 18(4): 567–588.

Colton, Craig. 2005. *An Unnatural Metropolis: Wrestling New Orleans from Nature*. Baton Rouge: Louisiana State University Press.

Colton, Craig, ed. 2000. *Transforming New Orleans and Its Environs: Centuries of Change*. Pittsburgh: University of Pittsburgh Press.

Comiso, J. C. 2002. A Rapidly Declining Perennial Sea Ice Cover in the Arctic. *Geophysical Research Letters* 29(20): 1956–1960.

Commission on Human Security. 2003. *Human Security Now*. New York: Commission on Human Security. Accessed May 25, 2007 <http://www.humansecuritychs.org/finalreport/English/FinalReport.pdf>.

Conca, Ken. 2002. Consumption and Environment in a Global Economy. In *Confronting Consumption*, ed. Thomas Princen, Michael Maniates, and Ken Conca, 133–153. Cambridge, MA: MIT Press.

———. 2006. *Governing Water. Contentious Transnational Politics and Global Institution Building*. Cambridge, MA: MIT Press.

Conca, Ken, and Geoffrey D. Dabelko. 2004. *Green Planet Blues. Environmental Politics from Stockholm to Johannesburg*, 3rd ed. Boulder, CO: Westview Press.

Conroy, Michael, and Amy K. Glasmeier. 1993. Unprecedented Disparities, Unparalleled Adjustment Needs: Winners and Losers on the NAFTA "Fast Track." *Journal of InterAmerican Studies and World Affairs* 34: 1–37.

Conway, Gordon, and Gary Toenniessen. 1999. Feeding the World in the Twenty-first Century. *Nature* 402 (Supplement): C55–C58.

Cox, Kevin. 1997. Introduction: Globalization and Its Politics in Question. In *Spaces of Globalization: Reasserting the Power of the Local*, ed. Kevin Cox, 1–19. New York: Guilford Press.

CPRC. 2004. *The Chronic Poverty Report, 2004–05*. Manchester, UK: Chronic Poverty Research Centre.

Crabbe, P., and M. Robin. 2006. Institutional Adaptation of Water Resource Infrastructures to Climate Change in Eastern Ontario. *Climatic Change* 78: 103–133.

Crate, Susan A. 2006. *Cows, Kin, and Globalization: An Ethnography of Sustainability.* Lanham, MD: AltaMira Press.

Crichton, Michael. 2004. *State of Fear.* New York: HarperCollins.

Cronon, William. 1991. *Nature's Metropolis: Chicago and the Great West.* New York: W.W. Norton.

Crosby, Alfred W., Jr. 1972. *The Columbian Exchange: Biological and Cultural Consequences of 1492.* Westport, CT: Greenwood Press.

Crowley, Sheila. 2006. Where Is Home? Housing for Low-Income People after the 2005 Hurricanes. In *There Is No Such Thing as a Natural Disaster: Race, Class and Hurricane Katrina*, ed. Chester Hartman and Gregory D. Squires, 121–166. New York: Routledge.

CSIRO (Commonwealth Scientific and Industrial Research Organisation)—Australia, Arizona State University—USA and Stockholm University—Sweden 2007. *Urban Resilience Research Prospectus.* Resilience Alliance. Accessed May 26, 2007 <http://www.resalliance.org/files/1172764197_urbanresilienceresearchprospectusv7feb07.pdf.>.

Csonka, Yvon, and Peter Schweitzer. 2004. Societies and Cultures: Change and Persistence. *AHDR (Arctic Human Development Report)*, 45–68. Akureyri: Stefansson Arctic Institute.

Curran, Lisa, S. N. Trigg, A. K. McDonald, D. Astiani, Y. M. Hardiono, P. Siregar, I. Caniago, and E. Kasischke. 2004. Lowland Forest Loss in Protected Areas of Indonesian Borneo. *Science* 303: 1000–1003.

Curtis, Fred. 2003. Eco-localism and Sustainability. *Ecological Economics* 46(1): 83–102.

Cutter, Susan. 2006. The Geography of Social Vulnerability: Race, Class, and Catastrophe. In Social Science Research Council, *Understanding Katrina: Perspectives from the Social Sciences.* Accessed June 15, 2007 <http://understandingkatrina.ssrc.org/Cutter/>.

Cutter, Susan L., Bryan J. Boruff, and W. Lynn Shirley. 2003. Social Vulnerability to Environmental Hazards. *Social Science Quarterly* 84(2): 242–261.

Cutter, Susan L., Christopher T. Emrich, Jerry T. Mitchell, Bryan J. Boruff, Melanie Gal, Mathew C. Schmidtlein, Christopher G. Burton, and Ginni Melton. 2006. The Long Road Home. Race, Class and Recovery from Hurricane Katrina. *Environment* 48(2): 8–20.

Datt, Gaurav, and Hans Hoogeveen. 2003. El Niño or El Peso? Crisis, Poverty and Income Distribution in the Philippines. *World Development* 31(7): 1103–1124.

Daviron, Benoit, and Stafano Ponte. 2005. *The Coffee Paradox: Global Markets, Commodity Trade and the Elusive Promise of Development.* London: Zed Books.

Davis, Donald W. 2000. Historical Perspective on Crevasses, Levees, and the Mississippi River. In *Transforming New Orleans and Its Environs: Centuries of Change*, ed. Craig Coulton, 84–106. Pittsburgh: University of Pittsburgh Press.

Davis, Mike. 2001. *Late Victorian Holocausts: El Nino Famines and the Making of the Third World.* London: Verso.

———. 2006. *Planet of Slums.* London: Verso.

Demeritt, David. 2001. The Construction of Global Warming and the Politics of Science. *Annals of the Association of American Geographers* 91(2): 307–337.

Denton, Fatma. 2002. Climate Change Vulnerability, Impacts, and Adaptation: Why Does Gender Matter? *Gender and Development* 10(2): 10–20.

Deshingkar, Priya, Usha Kulkarni, Laxman Rao, and Sreeniva Rao. 2003. Changing Food Systems in India: Resource-Sharing and Marketing Arrangements for Vegetable Production in Andra Pradesh. *Development Policy Review* 21(5–6): 627–639.

Devas, Nick. 2002. Urban Livelihoods—Issues for Urban Governance and Management. In *Urban Livelihoods: A People-Centred Approach to Reducing Poverty*, ed. Carole Rakodi and Tony Lloyd-Jones, 205–221. London: Earthscan.

Devereux, Stephen, and Simon Maxwell, eds. 2001. *Food Security in Sub-Saharan Africa*. Pietermaritzburg, South Africa: Institute of Development Studies.

De Wit, Maarten, and Jacek Stankiewicz. 2006. Changes in Surface Water Supply Across Africa with Predicted Climate Change. *Science* 311: 1917–1921.

DFID. 1999. *Sustainable Livelihood Fact Sheets: Introduction*. London: U.K. Department for International Development. Accessed June 15, 2007 <http://www.livelihoods.org/info/guidance_sheets_pdfs/section1.pdf>.

Diamond, Jared. 1999. *Guns, Germs, and Steel: The Fates of Human Societies*. New York: W.W. Norton.

———. 2005. *Collapse. How Societies Choose to Fail or to Succeed*. New York: Penguin Books.

Dicken, Peter. 2007. *Global Shift: Mapping the Changing Contours of the World Economy*, 5th ed. New York: The Guilford Press.

Dinar, Ariel, Robert Mendelsohn, Robert Evenson, Jyoti Parikh, Apurva Sanghi, Kavi Kumar, James McKinsey, and Stephen Lonergan. 1998. *Measuring the Impact of Climate Change on Indian Agriculture*. Washington, DC: World Bank Technical Paper, No. 402.

Dixon, Timothy H., Falk Amelung, Alessandro Ferritti, Fabrizio Novali, Fabio Rocca, Roy Dokka, Giovanni Sella, Sang-Wan Kim, Shimon Wdowinski, and Dean Whitman. 2006. Subsidence and Flooding in New Orleans. *Nature* 441.1 (June): 587–588.

DNV (Det Norske Veritas). 2003. *Evaluation of the Barents Sea as Particular Sensitive Sea Area*. Report No. 2002-1621. Høvik, Norway: Det Norske Veritas.

Dodds, Felix, and Tim Pippard, eds. 2005. *Human and Environmental Security: An Agenda for Change*. London: Earthscan.

Dollar, David, and Art Kraay. 2002. Spreading the Wealth. *Foreign Affairs* 81(1): 120–133.

———. 2004. Trade, Growth, and Poverty. *The Economic Journal* 114 (Feb.): F22–F29.

Dore, Mohammed H. I., and David Etkin. 2003. Natural Disasters, Adaptive Capacity and Development in the Twenty-first Century. In *Natural Disasters and Development in a Globalizing World*, ed. Mark Pelling, 75–91. London: Routledge.

Dow, Kirsten. 1992. Exploring Differences in our Common Futures(s): The Meaning of Vulnerability to Global Environmental Change. *Geoforum* 23: 417–436.

Dow, Kirstin, and Thomas E. Downing. 2006. *The Atlas of Climate Change: Mapping the World's Greatest Challenge*. Berkeley: University of California Press.

Dowdeswell, Julian A. 2006. The Greenland Ice Sheet and Global Sea-Level Rise. *Science* 311: 963–964.

Downing, Thomas E. 1991. Vulnerability to Hunger in Africa: A Climate Change Perspective. *Global Environmental Change* 1(5): 365–380.

Drèze, Jean, and Amartya Sen. 2002. *India: Development and Participation*, 2nd ed. Oxford: Oxford University Press.

Dryzek, John S. 2005. *The Politics of the Earth: Environmental Discourses*. Oxford: Oxford University Press.

Duhaime, Gérard. 2004. Economic Systems. In *AHDR (Arctic Human Development Report) (AHDR)*, 69–84. Akureyri, Iceland: Stefansson Arctic Institute.

Dunkley, Graham. 2004. *Free Trade. Myth, Reality and Alternatives*. London: Zed Books.

Dyson, Michael Eric. 2006. *Come Hell or High Water: Hurricane Katrina and the Color of Disaster*. New York: Basic Books.

Eakin, Hallie. 2003. "The Social Vulnerability of Irrigated Vegetable Farming Households in Central Puebla. *Journal of Environment and Development* 12(4): 414–429.

———. 2005. Institutional Change, Climate Risk and Rural Vulnerability: Cases from Central Mexico. *World Development* 33(11): 1923–1938.

———. 2006. *Weathering Risk in Rural Mexico: Climatic, Institutional, and Economic Change.* Tucson: The University of Arizona Press.

Eakin, Hallie, and Maria Carmen Lemos. 2006. Adaptation and the State: Latin America and the Challenge of Capacity-Building under Globalization. *Global Environmental Change* 16: 7–18.

Eakin, Hallie, and Amy Lynd Luers. 2006. Assessing the Vulnerability of Social-Environmental Systems. *Annual Review of Environment and Resources* 31: 365–394.

Eakin, Hallie, Catherine Tucker, and Edwin Castellanos. 2006. Responding to the Coffee Crisis: A Pilot Study of Farmers' Adaptations in Mexico, Guatemala and Honduras. *The Geographical Journal* 172: 156–171.

Earth Institute News. 2007. Researchers from around the World Converge on New York to Link Climate Change Science with Urban Policymaking Efforts. Columbia University, Earth Institute (posted May 7, 2007). Accessed May 31, 2007 <http://www.earthinstitute.columbia.edu/news/2007/story05-09-07.php>.

Eashvaraiah, P. 2001. Liberalisation, the State and Agriculture in India. *Journal of Contemporary Asia* 31(3): 331–346.

Easterling, William, and Colin Polsky. 2004. Crossing the Divide: Linking Global and Local Scales in Human-Environment Systems. In *Scale and Geographic Inquiry: Nature, Society, and Method*, ed. Eric Sheppard and Robert B. McMaster, 66–85. Malden, MA: Blackwell.

Easterling, W. E., P. K. Aggarwal, P. Batima, K. M. Brander, L. Erda, S. M. Howden, A. Kirilenko, J. Morton, J.-F. Soussana, J. Schmidhuber, and F. N. Tubiello. 2007. Food, Fibre and Forest Products. In *IPCC Fourth Assessment Report: Impacts, Vulnerability, and Adaptation*, ed. M. L. Parry, O. F. Canziani, J. P. Palutikof, P. J. van der Linden, and C. E. Hanson, 2273–2313. Cambridge: Cambridge University Press.

Eaton, Heather. 2003. Can Ecofeminism Withstand Corporate Globalization? In *Ecofeminism and Globalization: Exploring Culture, Context, and Religion*, ed. Heather Eaton and Lois Ann Lorentzen, 23–37. Lanham, MD: Rowman & Littlefield.

Eaton, Heather, and Lois Ann Lorentzen. 2003. Introduction. In *Ecofeminism and Globalization: Exploring Culture, Context, and Religion*, ed. Heather Eaton and Lois Ann Lorentzen, 1–7. Lanham, MD: Rowman & Littlefield.

eCensus India. 2002. Data on Workers and their Categories: An Insight. *eCensus India*, No. 7 (February 28, 2002). Accessed June 15, 2007 <http://www.censusindia.net/results/eci7_page3.html>.

EcoEquity and Christian Aid. 2006. *Greenhouse Development Rights: An Approach to the Global Climate Regime that Takes Climate Protection Seriously While Also Preserving the Right to Human Development.* Briefing Paper. Accessed March 7, 2007 <http://www.ecoequity.org/GDRs/GDRs_Nairobi.pdf>.

Economy Watch. 2007. Introduction. *Indian Economy Overview.* Accessed June 15, 2007 <http://www.economywatch.com/indianeconomy/indian-economy-overview.html>.

Ehrenreich, Barbara. 2001. *Nickel and Dimed: On (Not) Getting By in America.* New York: Metropolitan Books.

Ehrenreich, Barbara, and Arlie Russell Hochschild, eds. 2002. *Global Woman: Nannies, Maids, and Sex Workers in the New Economy.* New York: Henry Holt.

EJF. 2006. *Mangroves: Nature's Defence against Tsunamis—A Report on the Impact of Mangrove Loss and Shrimp Farm Development on Coastal Defences.* London: Environmental Justice Foundation.

Ellis, Frank. 2000. *Rural Livelihoods and Diversity in Developing Countries*. New York: Oxford University Press.

Elvin, Mark. 2004. *The Retreat of the Elephants: An Environmental History of China*. New Haven, CT: Yale University Press.

Emanuel, K., 2005. Increasing Destructiveness of Tropical Cyclones over the Past 30 Years. *Nature* 436, 686–688.

Energy Information Administration. 2006. *International Energy Outlook 2006*. Washington: U.S. Department of Energy.

Eriksen, Siri, and P. Mick Kelly. 2007. Developing Credible Vulnerability Indicators for Climate Adaptation Policy Assessment. *Mitigation and Adaptation Strategies for Global Change* 12(4): 495–524.

Eriksen, Siri, and Julie Silva. 2008. The Effect of Market Integration on Household Vulnerability to Climate Stress in Mozambique: Empirical Evidence of Multiple Stressors. *Environmental Science and Policy*. Forthcoming.

Escobar, Arturo. 1996. Constructing Nature: Elements for a Poststructural Political Ecology. In *Liberation Ecologies: Environment, Development, Social Movements*, ed. Richard Peet and Michael Watts, 46–68. London: Routledge.

Eskeland, Gunnar S., and Ann E. Harrison. 2003. Moving to Greener Pastures? Multinationals and the Pollution Haven Hypothesis. *Journal of Development Economics* 70: 1–23.

Esty, Daniel C. 1994. *Greening the GATT: Trade, Environment and the Future*. Washington, DC: International Institute for Economics.

Evans, Nick, Carol Morris, and Michael Winter. 2002. Conceptualizing Agriculture: A Critique of Post-productivism as the New Orthodoxy. *Progress in Human Geography* 26(3): 313–332.

FAO. 2003. *WTO Agreement on Agriculture: The Implementation Experience—Developing Country Case Studies*. Rome: Food and Agricultural Organization of the United Nations.

Farber, Daniel A. 2007. Basic Compensation for Victims of Climate Change. *University of Pennsylvania Law Review*. 155(6): 1605–1656.

Fearnside, Philip M. 2001. Soybean Cultivation as a Threat to the Environment in Brazil. *Environmental Conservation* 28(1): 23–38.

Ferlie, Ewan, Lynn Ashburner, Louise Fitzgerald, and Andrew Pettigrew. 1996. *The New Public Management in Action*. Oxford: Oxford University Press.

Fischer, Gunther, Harrij van Velthuizen, Mahendra Shah, and Freddy O. Nachtergaele. 2002. *Global Agro-ecological Assessment for Agriculture in the 21st Century: Methodology and Results*. Laxemburg, Austria: IIASA.

Fischetti, M. 2001. Drowning New Orleans. *Scientific American*, October: 77–85.

Foley, Jonathan A., Ruth DeFries, Gregory P. Asner, Carol Barford, Gordon Bonan, Stephen R. Carpenter, F. Stuart Chapin, Michael T. Coe, Gretchen C. Daily, Holly K. Gibbs, Joseph H. Helkowski, Tracey Holloway, Erica A. Howard, Christopher J. Kucharik, Chad Monfreda, Jonathan A. Patz, I. Colin Prentice, Navin Ramankutty, and Peter K. Snyder. 2005. Global Consequences of Land Use. *Science* 309: 570–574.

Folke, Carl. 2006. Resilience: The Emergence of a Perspective for Social–ecological Systems Analyses. *Global Environmental Change* 16: 253–267.

Folke, Carl, Steve Carpenter, Thomas Elmqvist, Lance Gunderson, C. S. Holling, and Brian Walker. 2002. Resilience and Sustainable Development: Building Adaptive Capacity in a World of Transformations. *Ambio* 31(5): 437–440.

Ford, James D., Barry Smit, and Johanna Wandel. 2006. Vulnerability to Climate Change in the Arctic: A Case Study from Arctic Bay, Canada. *Global Environmental Change* 16(2): 145–160.

Forsyth, Tim. 2003. *Critical Political Ecology: The Politics of Environmental Science*. London: Routledge.

Foucault, Michel. 1977. *Discipline and Punish: The Birth of the Prison* (Alan M. Sheridan-Smith trans). New York: Pantheon Books.

Frankenberg, Elizabeth, James P. Smith, and Duncan Thomas. 2002. *Economic Shocks, Wealth and Welfare*. Seattle: University of Washington, Center for Research on Families, Distinguished Speaker Series: 2001–2002. Accessed June 15, 2007 <http://depts.washington.edu/crfam/seminarseries01-02/Thomas.pdf>.

Friedman, Thomas. 2000. *The Lexus and the Olive Tree: Understanding Globalization*. New York: Anchor.

———. 2005. *The World Is Flat: A Brief History of the Twenty-First Century*. New York: Farrar, Straus and Giroux.

Fuhrer, Jürg. 2003. Review. Agroecosystem Responses to Combinations of Elevated CO_2, Ozone and Global Climate Change. *Agriculture, Ecosystems and Environment* 97(1–3): 1–20.

Funke, Nikki, Karen Nortje, Kieran Findlater, Mike Burns, Anthony Turton, Alex Weaver, and Hanlie Hattingh. 2007. Redressing Inequality: South Africa's New Water Policy. *Environment* 49(3): 10–23.

Gadgil, Sulochana. 1995. Climate Change and Agriculture: An Indian Perspective. *Current Science,* 69(8): 649–659.

Gan, Lin. 2003. Globalization of the Automobile Industry in China: Dynamics and Barriers in Greening of the Road Transportation. *Energy Policy* 31(6): 537–551.

García-Acevedo, María Rosa, and Helen Ingram. 2004. Conflict in the Borderlands. *NACLA Report on the Americas* 38(1): 19–24.

Gasper, Des. 2005. Securing Humanity: Situating Human Security as Concept and Discourse. *Journal of Human Development* 7(2): 221–245.

GECHS 1999. *Science Plan: Global Environmental Change and Human Security Project*. Bonn: International Human Dimensions Programme (IHDP). Accessed May 26, 2007 <http://www.ihdp.uni-bonn.de/html/publications/reports/report11/gehssp.htm>.

GEF. 2006. *Programming Paper for Funding the Implementation of NAPAs under the LDC Trust Fund*. United Nations, Global Environment Facility, GEF/C.28/18. Accessed January 16, 2007 <http://www.thegef.org/Documents/Council_Documents/GEF_C28/documents/C.28.18LDCTrustFund_000.pdf>.

Gilbert, Emily. 2005. The Inevitability of Integration? Neoliberal Discourse and the Proposals for a New North American Economic Space after September 11. *Annals of the Association of American Geographers* 95(1): 202–222.

Girot, Pascal O. 2002. Environmental Degradation and Regional Vulnerability: Lessons from Hurricane Mitch. In *Conserving the Peace: Resources, Livelihoods and Security*, ed. Richard Matthew, Mark Halle, and Jason Switzer, 273–317. Winnipeg: International Institute for Sustainable Development (IISD).

Glantz, Michael H. 1995. Assessing the Impacts of Climate: The Issue of Winners and Losers in a Global Climate Change Context. In *Climate Change Research: Evaluation and Policy Implications*, ed. S. Zwerver, R. S. A. R. van Rompaey, M. T. J. Kok, and M. M. Berk, 41–54. Amsterdam: Elsevier Science.

Glasmeier, Amy K. 2005. *An Atlas of Poverty in America: One Nation Pulling Apart, 1960–2003*. New York: Routledge.

Gleeson, Brendan, and Nicholas Low. 2000. Cities as Consumers of the World's Environment. In *Consuming Cities: The Urban Environment in the Global Economy after the Rio Declaration*, ed. Nicholas Low, Brendan Gleeson, Ingemar Elander, and Rolf Lidskog, 1–29. London: Routledge.

Gleick, Peter H., Gary Wolf, Elizabeth L. Chalecki, and Rachel Reyes, 2002. *The New Economy of Water: The Risks and Benefits of Globalization and Privatization of Fresh Water*. Oakland: Pacific Institute for Studies in Development, Environment, and Security.

Godschalk, David R. 2003. Urban Hazard Mitigation: Creating Resilient Cities. *Natural Hazards Review* 4(3): 136–143.

Goldman, Erica. 2002. Even in the High Arctic, Nothing Is Permanent. *Science* 297(1493a): doi: 10.1126/science.297.5586.1493a.

Goodman, David, and Michael Watts, eds. 1997. *Globalising Food: Agrarian Questions and Global Restructuring,* 1st ed. London: Routledge.

Gotham, Kevin Fox. 2002. Marketing Mardi Gras: Commodification, Spectacle and the Political Economy of Tourism in New Orleans. *Urban Studies* 39(10): 1735–1756.

Granberg, Alexander G. 1998. The Northern Sea Route: Trends and Prospects of Commercial Use. *Ocean & Coastal Management* 41(2–3): 175–207.

Greenpeace. 2007. Greenpeace Briefing: Carbon Capture and Storage. Greenpeace International. Accessed May 26 2007 <http://www.greenpeace.org/raw/content/international/press/reports/technical-brifing-ccs.pdf>.

Greider, William. 1997. *One World, Ready or Not: The Manic Logic of Global Capitalism*. New York: Touchstone Books.

Gu, Chaolin, and Jianafa Shen. 2003. Transformation of Urban Socio-Spatial Structure in Socialist Market Economies: The Case of Beijing. *Habitat International* 27: 107–122.

Guimarães, Roberto P. 2004. Waiting for Godot: Sustainable Development, International Trade and Governance in Environmental Policies. *Contemporary Politics* 10(3–4): 203–225.

Gulati, Ashok. 2002. Indian Agriculture in a Globalizing World. *American Journal of Agricultural Economics* 84(3): 754–761.

Gulati, Ashok, and Tim Kelley. 1999. *Trade Liberalization and Indian Agriculture: Cropping Pattern Changes and Efficiency Gains in Semi-Arid Tropics*. New Delhi: Oxford University Press.

Gullette, Margaret Morganroth. 2006. Katrina and the Politics of Later Life. In *There Is No Such Thing as a Natural Disaster: Race, Class and Hurricane Katrina*, ed. Chester Hartman and Gregory D. Squires, 103–120. New York: Routledge.

Gunderson, Lance, and C. S. Holling, eds. 2002. *Panarchy: Understanding Transformations in Human and Natural Systems*. Washington, DC: Island Press.

Gupta, Sujata, Akram Javed, and Divya Datt. 2003. Economics of Flood Protection in India. *Natural Hazards* 28: 199–210.

Haas, Peter M., ed. 2003. *Environment in the New Global Economy*. Cheltenham, UK: Edward Elgar Publishing.

HABITAT (United Nations Centre for Human Settlements). 2001. *Cities in a Globalizing World, Global Report on Human Settlements 2001*. Sterling, VA: Earthscan.

Halweil, Brian. 2000. Where Have All the Farmers Gone? *World Watch*, September/October 2000.

Hanson, James E. 1988. *The Greenhouse Effect: Impacts on Current Global Temperature and Regional Heat Waves*. Statement presented to the U.S. Senate Committee on Energy and Natural Resources. Washington, DC, June 23.

Hardoy, Jorge E., Diana Mitlin, and David Satterthwaite. 2001. *Environmental Problems in an Urbanizing World*. London: Earthscan.

Hardt, Michael, and Antonio Negri. 2004. *Multitude: War and Democracy in the Age of Empire*. New York: Penguin Press.

Hartman, Chester, and Gregory Squires. 2006. Pre-Katrina, Post-Katrina. In *There Is No Such Thing as a Natural Disaster: Race, Class and Hurricane Katrina*, ed. Chester Hartman and Gregory D. Squires, 1–12. New York: Routledge.

Harvey, David. 1990. *The Condition of Post-Modernity: An Enquiry into the Origins of Cultural Change*. Malden, MA: Blackwell.

Hecht, Susanna B. 2005. Soybeans, Development and Conservation on the Amazon Frontier. *Development and Change* 36(2): 375–404.

Hecht, Susanna B., Susan Kandel, Eliana Gomes, Nelson Cuellar, and Herman Rosa. 2006. Globalization, Forest Resurgence, and Environmental Politics in El Salvador. *World Development* 34(2): 308–323.

Heininen, Lasse. 2004. Circumpolar International Relations and Geopolitics. In *AHDR (Arctic Human Development Report)*, 207–225. Akureyri, Iceland: Stefansson Arctic Institute.

Held, David. 2004. *Global Covenant: The Social Democratic Alternative to the Washington Consensus*. Cambridge, UK: Polity Press.

Held, David, and Ayse Kaya, eds. 2007. *Global Inequality*. Cambridge, UK: Polity Press.

Held, David, and Anthony McGrew. 2002. *Globalization/Anti-Globalization*. Cambridge, UK: Polity Press.

———, eds. 2000. *The Global Transformations Reader: An Introduction to the Globalization Debate*. Cambridge, UK: Polity Press.

Held, David, Anthony McGrew, David Goldblatt, and Jonathan Perraton. 1999. *Global Transformations: Politics, Economics and Culture*. Cambridge, UK: Polity Press.

Helleiner, Eric. 2002. Think Globally, Transact Locally: The Local Currency Movement and Green Political Economy. In *Confronting Consumption*, ed. Thomas Princen, Michael Maniates, and Ken Conca, 255–273. Cambridge, MA: MIT Press.

Hertel, Thomas W., Maros Ivanic, Paul V. Preckel, John A. L. Cranfield, and Will Martin. 2003. Short- Versus Long-Run Implications of Trade Liberalization for Poverty in Three Developing Countries. *American Journal of Agricultural Economics* 85(5): 1299–1306.

Hertel, Thomas W., and L. Alan Winters. 2006. *Poverty and the WTO: Impacts of the Doha Development Agenda*. Washington, DC: Palgrave MacMillan and World Bank.

Hewitt, Kenneth. 1997. *Regions of Risk, A Geographical Introduction to Disasters*. Harlow, Essex: Longman.

Hilhorst, Dorothea. 2004. Complexity and Diversity: Unlocking Social Domains of Disaster Response. In *Mapping Vulnerability: Disasters, Development & People*, ed. Greg Bankoff, Georg Frerks, and Dorothea Hilhorst, 52–66. London: Earthscan.

Hines, Colin. 2003. Time to Replace Globalization with Localization. *Global Environmental Politics* 3(3): 1–7.

Hirst, Paul, and Grahame Thompson. 1999. *Globalization in Question: The International Economy and the Possibilities of Governance,* 1st ed. Malden, MA: Polity Press.

Holland, Geoff. 2002. The Arctic Ocean—The Management of Change in the Northern Seas. *Ocean & Coastal Management* 41(11–12): 841–851.

Holland, Marika M., Cecilia M. Bitz, and Bruno Tremblay. 2006. Future Abrupt Reductions in the Summer Arctic Sea Ice. *Geophysical Research Letters* 33(23): L23504, doi: 10.1029/2006GL028024.

Holling, C. S. 1973. Resilience and Stability of Ecological Systems. *Annual Review of Ecology and Systematics* 4: 1–23.

———. 2004. From Complex Regions to Complex Worlds. *Ecology and Society* 9(1): 11.

Homer-Dixon, Thomas. 2006. *The Upside of Down: Catastrophe, Creativity, and the Renewal of Civilization*. Washington, DC: Island Press.

Hoogvelt, Ankie. 1997. *Globalization and the Postcolonial World*. Baltimore: Johns Hopkins Press.

Horne, Jed. 2006. *Breach of Faith: Hurricane Katrina and the Near Death of a Great American City*. New York: Random House.

Hulme, Mike, ed. 1996, *Climate Change and Southern Africa: An Exploration of Some Potential Impacts and Implications in the SADC Region*. Norwich, UK: WWF International and Climate Research Unit, UEA.

Huq, Saleemul, and Ian Burton. 2003. Funding Adaptation to Climate Change: What, Who and How to Fund? *Sustainable Development Opinion Papers*. London: IIED.

Huynen, M. M. T. E, P. Martens, and H. B. M. Hilderlink. 2005. *The Health Impacts of Globalisation: A Conceptual Framework*. Report 550012007/2005, Netherlands Environmental Assessment Agency. Accessed May 26, 2007 <http://www.mnp.nl/bibliotheek/rapporten/550012007.pdf>.

ICLEI (International Council for Local Environmental Initiatives). 2006. Cities for Climate Protection. 2006. Accessed January 11, 2007 <http://www.iclei.org/index.php?id=800>.

IFAD (International Fund for Agricultural Development). 2001. *Rural Poverty Report 2001: The Challenge of Ending Rural Poverty*. New York: Oxford University Press.

IFPRI. 2003. A Level Playing Field for Poor Farmers. *IFPRI Forum* (March 2003). Washington, DC: International Food Policy Research Institute.

Ikeme, Jekwu. 2003. Equity, Environmental Justice and Sustainability: Incomplete Approaches in Climate Change Politics. *Global Environmental Change* 13: 195–206.

Ingram, Jane C., Guillermo Franco, Cristina Rumbaitis-del Rio, and Bjian Khazai. 2006. Post-disaster Recovery Dilemmas: Challenges in Balancing Short-term and Long-term Needs for Vulnerability Reduction. *Environmental Science & Policy* 9(7–8): 607–613.

INSROP. 1998. *Environmental Conditions Affecting Commercial Shipping*. Final Report, INSROP Phase 2 Projects: Lysaker: The INSROP Secretariat. Accessed May 26, 2007 <http://www.ims.uaf.edu/insrop-2/report.html>.

International Energy Agency. 2006. *World Energy Outlook 2006*. Paris: International Energy Agency. Accessed May 27, 2007 <http://www.worldenergyoutlook.org/2006.asp>.

International Forum on Globalization. 2002. *Alternatives to Economic Globalization: A Better World Is Possible*. San Francisco: Berrett-Koehler Publishers.

Inuit Circumpolar Conference. 2005. *Petition to the Inter American Commission on Human Rights Seeking Relief from Violations Resulting from Global Warming Caused by Acts and Omissions of the United States*. Iqaluit, Nunavut, Canada. Accessed May 29, 2007 <http://www.ciel.org/Publications/ICC_Petition_7Dec05.pdf >.

Ionescu, Cezar, Richard J. T. Klein, Jochen Hinkel, K. S. Kavi Kumar, and Rupert Klein. 2005. *Towards a Formal Framework of Vulnerability to Climate Change*. Potsdam, Germany, PIK.

IPCC. 2007a. *Climate Change 2007: Impacts, Adaptation and Vulnerability. Contribution of Working Group II to the Fourth Assessment Report of the Intergovernmental Panel on Climate Change* [M. L. Parry, O. F. Canziani, J. P. Palutikof, P. J. van der Linden, and C. E. Hanson, eds.]. Cambridge: Cambridge University Press,

IPCC. 2007b. *Climate Change 2007: Mitigation. Contribution of Working Group III to the Fourth Assessment Report of the Intergovernmental Panel on Climate Change* [B. Metz, O. R. Davidson, P. R. Bosch, R. Dave, and L. A. Meyer, eds.]. Cambridge: Cambridge University Press.

IPCC. 2007c. *Climate Change 2007: The Physical Science Basis. Contribution of Working Group I to the Fourth Assessment Report of the Intergovernmental Panel on Climate Change* [S. Solomon, D. Qin, M. Manning, Z. Chen, M. Marquis, K. B. Averyt, M. Tignor, and H. L. Miller, eds.]. Cambridge: Cambridge University Press.

IUCN. 2006. *2006 IUCN Red List of Threatened Species*. The World Conservation Union (International Union for the Conservation of Nature and Natural Resources). Accessed June 12, 2007 <http://www.iucnredlist.org/>.

Jacques, Peter. 2006. The Rearguard of Modernity: Environmental Skepticism as a Struggle of Citizenship. *Global Environmental Politics* 6(1): 76–101.

Jha, Raghbendra. 2004. Reducing Poverty and Inequality in India: Has Liberalization Helped? In *Inequality Growth and Poverty in an Era of Liberalization and Globalization*, ed. Giovanni Andrea Cornia, 297–327. New York: Oxford University Press.

Jodha, Narpat S. 2000. Globalization and Fragile Mountain Environments: Policy Challenges and Choices. *Mountain Research and Development* 20(4): 296–299.

Johannessen, Ola M., Elena V. Shalina, and Martin W. Miles. 1999. Satellite Evidence for an Arctic Sea Ice Cover in Transformation. *Science* 286: 1937–1939.

Johnson, Pierre-Marc, and Andre Beaulieu, eds. 1996. The Environment and NAFTA: Understanding and Implementing the New Continental Law. Washington, DC: Island Press.

Johnston, Paul, and David Santillo. 2002. Carbon Capture and Sequestration: Potential Environmental Impacts. In *Proceedings from IPCC Workshop on Carbon Dioxide Capture and Storage*, 95–100. Regina, Canada, November 19–21, 2002.

Jones, Peris, and Kristian Stokke. 2005. Introduction—Democratising Development: The Politics of Socio-Economic Rights. In *Democratising Development: The Politics of Socio-Economic Rights in South Africa*, ed. Peris Jones and Kristian Stokke, 1–38. Leiden: Martinus Nijhoff Publishers.

Jones, Peter G., and Philip K. Thornton. 2003. The Potential Impacts of Climate Change on Maize Production in Africa and Latin America in 2055. *Global Environmental Change* 13 (1): 51–59.

Jones, Roger N. 2001. An Environmental Risk Assessment/Management Framework for Climate Change Impact Assessments. *Natural Hazards* 23: 197–230.

Jones, Samantha. 2002. Social Constructionism and the Environment: Through the Quagmire. *Global Environmental Change* 12: 247–251.

Jones, Tom. 2002. Globalization and Environmental Sustainability: An OECD Perspective. *International Journal of Sustainable Development* 3(2): 146–158.

Jorgensen, Andrew K. 2007. Foreign Direct Investment and Pesticide Use Intensity in Less-Developed Countries: A Quantitative Investigation. *Society & Natural Resources* 20: 73–83.

Kapstein, Ethan. 2000. Winners and Losers in the Global Economy. *International Organization* 54(2): 359–384.

Karl, Thomas R., and Kevin E. Trenberth. 2003. Modern Global Climate Change. *Science* 302: 1719–1723.

Kasperson, Roger E., Jeanne X. Kasperson, and Kirsten Dow. 2001. Vulnerability, Equity, and Environmental changes. In *Global Environmental Risk*, ed. Jeanne X. Kasperson and Roger E. Kasperson, 247–272. Tokyo: United Nations University Press and Earthscan.

Kates, Robert W., and Ian Burton, eds. 1986. *Geography, Resources, and Environment, Volume 1: Selected Writings of Gilbert F. White*. Chicago: University of Chicago Press.

Kates, Robert W., B. L. Turner II, and William C. Clark. 1990. The Great Transformation. In *The Earth as Transformed by Human Action: Global and Regional Changes in the Biosphere over the Past 300 Years*, ed. B. L. Turner II, William C. Clark, Robert W. Kates, John F. Richards, Jessica T. Mathews, and William B. Meyer, 1–17. Cambridge: Cambridge University Press with Clark University.

Kates, Robert W., William C. Clark, Robert Corell, J. Michael Hall, Carlo C. Jaeger, Ian Lowe, James J. McCarthy, Hans Joachim Schellnhuber, Bert Bolin, Nancy M. Dickson, Sylvie Faucheux, Gilberto C. Gallopin, Arnulf Grübler, Brian Huntley, Jill Jäger, Narpat S. Jodha, Roger E. Kasperson, Akin Mabogunje, Pamela Matson, Harold Mooney, Berrien Moore III, Timothy O'Riordan, and Uno Svedlin. 2001. Sustainability Science. *Science* 292: 641–642.

Kay, John. 2004. *The Truth about Markets: Why Some Nations Are Rich but Most Remain Poor.* London: Penguin Books.

Kearney, A. T. 2005. Measuring Globalization. *Foreign Policy*, May/June 2005. Accessed June 12, 2007 <http://www.atkearney.com/shared_res/pdf/2005G-index.pdf>.

Kearney, Michael. 1984. *World View.* Novato, CA: Chandler and Sharp.

Keatinge, William R. 2003. Death in Heat Waves. *BMJ* 327: 512–513.

Kemp, René, and Saeed Parto. 2005. Governance for Sustainable Development: Moving from Theory to Practice. *International Journal of Sustainable Development* 8: 12–30.

Kennedy, P. Lynn, and Won W. Koo, eds. 2002. *Agricultural Trade Policies in the New Millennium.* New York: Food Products Press.

Khandlhela, Masingita, and Julian May. 2006. Poverty, Vulnerability and the Impact of Flooding in the Limpopo Province, South Africa. *Natural Hazards* 39: 275–287.

Khor, Martin. 2001. *Rethinking Globalization: Critical Issues and Policy Choices.* London: Zed Books.

Khvochtchinski, Nikolay I., and Yuriy M. Batskikh. 1998. The Northern Sea Route as an Element of the ICZM System in the Arctic: Problems and Perspectives. *Ocean & Coastal Management* 41(2–3): 161–173.

Kindleberger, Charles P. 1996. *World Economic Primacy: 1500 to 1990.* New York: Oxford University Press.

Kineman, John J. 1991. Gaia: Hypothesis or Worldview? In *Scientists on Gaia*, ed. Stephen H. Schneider and Penelope J. Boston, 47–65. Cambridge, MA: MIT Press.

Kingsnorth, Paul. 2003. *One No, Many Yeses: A Journey to the Heart of the Global Resistance Movement.* London: Free Press.

Kirby, Andrew. 2004. The Global Cultural Factory. In *Globalization and Its Outcomes,* ed. John O'Loughlin, Lynn Staeheli, and Edward Greenberg, 133–155. New York: Guilford Press.

Kirby, Peadar. 2006. *Vulnerability and Violence: The Impact of Globalization.* London: Pluto Press.

Klein, Naomi. 2000. *No Logo.* London: Harper Collins Publishers.

Klein, Richard J. T., and Robert J. Nicholls. 1999. Assessment of Coastal Vulnerability to Climate Change. *Ambio* 28(2): 182–187.

Klein, Richard J. T., Robert J. Nicholls, and Frank Thomalla. 2003. Resilience to Natural Hazards: How Useful Is This Concept? *Environmental Hazards* 5: 25–45.

Klinenberg, Eric. 2002. *Heat Wave: A Social Autopsy of Disaster in Chicago.* Chicago: University of Chicago Press.

Klooster, Dan. 2003. Forest Transitions in Mexico: Institutions and Forests in a Globalized Countryside. *The Professional Geographer* 55(2): 227–237.

Knowles, Richard D. 2006. Transport Shaping Space: Differential Collapse in Time-Space. *Journal of Transport Geography* 14: 407–425.

Kodoth, Praveena, and Mridul Eapen. 2005. Looking beyond Gender Parity: Gender Inequities of Some Dimensions of Well-Being in Kerala. *Economic and Political Weekly*, July 23, 2005: 3278–3286.

Koivusalo, Meri. 2006. On Equity, Inequality, and Global Institutions. *Global Social Policy* 6(2): 221–224.

Koocheki, A., and S. R. Gliessman. 2005. Pastoral Nomadism, a Sustainable System for Grazing Land Management in Arid Areas. *Journal of Sustainable Agriculture* 25(4): 113–131.

Kousky, Carolyn, and Stephen H. Schneider. 2003. Global Climate Policy: Will Cities Lead the Way? *Climate Policy* 3: 359–372.

Kraas, Frauke. 2003. Megacities as Global Risk Areas. *Petermanns Geographische Mitteilungen* 147(4): 6–15.

Kruger, Joseph A., and William A. Pizer. 2004. Greenhouse Gas Trading in Europe: The New Grand Policy Experiment. *Environment* 46(8): 8–23.

Krupnik, Igor, and Dyanna Jolly, eds. 2002. *The Earth Is Faster Now: Indigenous Observations of Arctic Environmental Change.* Fairbanks, AK: Arctic Research Consortium of the United States.

Kulkarni, Anil V., I. M. Bahuguna, B. P. Rathore, S. K. Singh, S. S. Randhawa, R. K. Sood, and Sunil Dhar. 2007. Glacial Retreat in Himalaya Using Indian Remote Sensing Satellite Data. *Current Science* 92(1): 69–74.

Kumar, K. S. K., and J. Parikh. 2001. Indian Agriculture and Climate Sensitivity. *Global Environmental Change* 11: 147–154.

Kumar, K. R., K. K. Kumar, R. G. Ashrit, S. K. Patwardhan, and G. B. Pant. 2002. Climate Change in India: Observations and Model Projections. In *Climate Change and India: Issues, Concerns and Opportunities*, ed. P. R. Shukla, K. S. Subodh, and P. V. Ramana, 24–75. New Delhi: Tata McGraw-Hill Publishing Company.

Kydd, Jonathan. 2002. *Agriculture and Rural Livelihoods: Is Globalization Opening or Blocking Paths out of Rural Poverty?* Agricultural Research and Extension Network Paper No. 121. Sted: Overseas Development Institute, UK. Accessed June 15, 2007 <http://www.odi.org.uk/agren/papers/agrenpaper_121.pdf>.

Lagadec, Patrick. 2004. Understanding the French 2003 Heat Wave Experience: Beyond the Heat, a Multi-Layered Challenge. *Journal of Contingencies and Crisis Management* 12(4): 160–169.

Laidler, Gita J. 2006. Inuit and Scientific Perspectives on the Relationship between Sea Ice and Climate Change: The Ideal Complement? *Climatic Change* 78(2–4): 407–444.

Lal, M., K. K. Singh, L. S. Rathore, G. Srinivasan, and S. A. Saseendran. 1998. Vulnerability of Rice and Wheat Yields in North-west India to Future Changes in Climate. *Agricultural and Forest Meteorology* 89: 101–114.

Lambin, Eric F., B. L. Turner, Helmut J. Geist, Samuel B. Agbola, Arild Angelsen, John W. Bruce, Oliver T. Coomes, Rodolfo Dirzo, Günther Fischer, Carl Folke, P. S. George, Katherine Homewood, Jacques Imbernon, Rik Leemans, Xiubin Li, Emilio F. Moran, Michael Mortimore, P. S. Ramakrishnan, John F. Richards, Helle Skånes, Will Steffen, Glenn D. Stone, Uno Svedin, Tom A. Veldkamp, Coleen Vogel, and Jianchu Xu. 2001. The Causes of Land-Use and Land-Cover Change: Moving beyond the Myths. *Global Environmental Change* 11(4): 261–269.

Lankao, Patricia Romero. 2007. Are We Missing the Point? Particularities of Urbanization, Sustainability and Carbon Emissions in Latin American Cities. *Environment and Urbanization* 19(1): 159–175.

Laurance, William F., Mark A. Cochrane, Scott Bergen, Philip M. Fearnside, Patricia Delamônica, Christopher Barber, Sammya D'Angelo, and Tito Fernandes. 2001. The Future of the Brazilian Amazon. *Science* 291(5503): 438–439.

Laxon, S., N. Peackock, and D. Smith. 2003. High Interannual Variability of Sea Ice Thickness in the Arctic Region. *Nature* 425: 947–950.

Lee, Kai N. 1999. Appraising Adaptive Management. *Conservation Ecology* 3(2): 3. Accessed June 15, 2007 <http://www.consecol.org/vol3/iss2/art3/>.

Lee, Yok-shiu F. 2007. Motorization in Rapidly Developing Cities. In *Scaling Urban Environmental Challenges: From Local to Global and Back*, ed. Peter Marcotullio and Gordon McGranahan, 179–205. London: Earthscan.

Lee, Yong-Sook, and Brenda S. A. Yeoh. 2004. Introduction: Globalisation and the Politics of Forgetting. *Urban Studies* 41: 2295–2301.

Leichenko, Robin, and Karen O'Brien. 2002. The Dynamics of Rural Vulnerability to Global Change: The Case of Southern Africa. *Mitigation and Adaptation Strategies for Global Change* 7: 1–18.

———. 2006. Is It Appropriate to Identify Winners and Losers? In *Fairness in Adaptation to Climate Change*, ed. W. Neil Adger, Jouni Paavola, Saleem Huq, and M. J. Mace, 96–114. Cambridge, MA: MIT Press.

Leichenko, Robin M., and Julie A. Silva. 2004. International Trade, Employment, and Earnings: Evidence from U.S. Rural Counties. *Regional Studies* 38(4): 353–372.

Leichenko, Robin M., and William D. Solecki. 2005. Exporting the American Dream: The Globalization of Suburban Consumption Landscapes. *Regional Studies* 39(2): 241–253.

———. 2006. Global Cities and Local Vulnerabilties. *IHDP Update* 2/06: 10–12.

———. 2008. Consumption, Inequity, and Environmental Justice: The Making of New Metropolitan Landscapes in Developing Countries. *Society and Natural Resources*. Forthcoming.

Lemos, Maria Carmen, and Arun Agrawal. 2006. Environmental Governance. *Annual Review of Environment and Resources* 31: 297–325.

Le Premier Ministre de la France. 2006. Plan "Canicule" 2006. Accessed March 2, 2007 <http://www.premierministre.gouv.fr/information/actualites_20/plan_canicule_2006_56310.html>.

Lind, Jeremy, and Siri Eriksen. 2006. The Impacts of Conflict on Household Coping Strategies: Evidence from Turkana and Kitui Districts in Kenya. *Die Erde* 137: 249–270.

Lindzen, Richard S. 2006. There Is No 'Consensus' on Global Warming, *Wall Street Journal*, June 26, 2006, A14.

Linerooth-Bayer, Joanne, and Anna Vári. 2006. Extreme Weather and Burden Sharing in Hungary. In *Fairness in Adaptation to Climate Change*, ed. W. Neil Adger, Jouni Paavola, Saleem Huq, and M. J. Mace, 239–259. Cambridge, MA: MIT Press.

Liotta, Peter H., and Allan W. Shearer. 2006. *Gaia's Revenge: Climate Change and Humanity's Loss*. Westport, CT: Praeger Publishers.

Lister, Ruth. 2004. *Poverty*. Cornwall, UK: Polity Press.

Litchfield, Julie, Neil McCulloch, and L. Alan Winters. 2003. Agricultural Trade Liberalization and Poverty Dynamics in Three Developing Countries. *American Journal of Agricultural Economics* 85(5): 1285–1291.

Liu, Amy, Mia Mabanta, and Matt Fellowes. 2006. *Katrina Index: Tracking Variables of Post-Katrina Recovery*. Washington, DC: The Brookings Institution, Metropolitan Policy Program.

Liverman, Diana M. 1990. Drought Impacts in Mexico: Climate, Agriculture, Technology, and Land Tenure in Sonora and Puebla. *Annals of the American Association of Geographers* 80: 49–72.

———. 2004. Who Governs, At What Scale and At What Price? Geography, Environmental Governance and the Commodification of Nature. *Annals of the Association of American Geographers* 94(4): 734–738.

Liverman, Diana M., and Silvina Vilas. 2006. Neoliberalism and the Environment in Latin America. *Annual Review of Environment and Resources* 31: 2.1–2.37.

Livingstone, David N. 1992. *The Geographical Tradition.* Oxford: Blackwell.

Logan, John. 2006. *The Impact of Katrina: Race and Class in Storm Damaged Neighborhoods.* Working Paper. Providence, RI: Brown University, Spatial Structures in the Social Sciences.

Lohmann, Larry. 2006. Carbon Trading: A Critical Conversation on Climate Change, Privatisation, and Power. *Development Dialogue* 48, Uppsala: Dag Hammerskjold Foundation.

Lomborg, Bjørn. 2001. *The Skeptical Environmentalist. Measuring the Real State of the World.* Cambridge, UK: Cambridge University Press.

Louisiana Department of Health and Hospitals. 2006. *2006 Louisiana Health and Population Survey, Survey Report, November 28, 2006: Orleans Parish.* Baton Rouge.

Lovelock, James. 1988. *The Ages of Gaia: A Biography of Our Living Earth.* New York: Bantam Books.

———. 2006. *The Revenge of Gaia: Earth's Climate Crisis and the Fate of Humanity.* London: Allen Lane.

Lubchenko, Jane. 2003. Waves of the Future: Sea Changes for a Sustainable World. In *Worlds Apart: Globalization and the Environment*, ed. James Gustave Speth, 21–31. Washington, DC: Island Press.

Luers, Amy L. 2005. The Surface of Vulnerability: An Analytical Framework for Examining Environmental Change. *Global Environmental Change* 15, 214–223.

Luers, A. L., D. B. Lobell, L. S. Sklar, C. L. Addams, and P. A. Matson. 2003. A Method for Quantifying Vulnerability, Applied to the Agricultural System of the Yaqui Valley, Mexico. *Global Environmental Change* 13: 255–267.

Luo, Q., and E. Lin. 1999. Agricultural Vulnerability and Adaptation in Developing Countries: The Asia-Pacific Region. *Climatic Change* 43: 729–743.

Lynas, Mark. 2003. *High Tide: News from a Warming World.* London: Flamingo.

Magnaghi, Alberto. 2000. *The Urban Village: A Charter for Democracy and Local Self-Sustainable Development.* London: Zed Books.

Mall, R. K., Ranjeet Singh, Akhilesh Gupta, G. Srinivasan, and L. S. Rathore. 2006. Impact of Climate Change on Indian Agriculture: A Review. *Climatic Change* 78(2–4): 445–478.

Mander, Jerry. 2003. Intrinsic Negative Effects of Economic Globalization on the Environment. In *Worlds Apart: Globalization and the Environment*, ed. James Gustave Speth, 109–130. Washington, DC: Island Press.

Marcotullio, Peter. 2001. Asian Urban Sustainability in the Era of Globalization. *Habitat International* 25(4): 577–598.

Marcotullio, Peter J., and Yok-Shiu F. Lee. 2003. Urban Environmental Transitions and Urban Transportation Systems. *International Development Planning Review* 25: 325–354.

Marsden, Terry, Gavin Bridge, and Phil McManus. 2002. Beyond the Social Construction of Nature: Rethinking Political Economies of the Environment. *Journal of Environmental Policy & Planning* 4: 103–105.

Martin, Patricia M. 2005. Comparative Topographies of Neoliberalism in Mexico. *Environment and Planning A*, 37: 203–220.

McCarthy, J. J., O. F. Canziani, N. A. Leary, D. J. Dokken, and K. S. White, eds. 2001. *Climate Change 2001: Impacts, Adaptation & Vulnerability Contribution of Working Group II to the Third Assessment Report of the Intergovernmental Panel on Climate Change* (IPCC). Cambridge: Cambridge University Press.

McCarthy, James. 2004. Privatizing Conditions of Production: Trade Agreements as Neoliberal Environmental Governance. *Geoforum* 35: 327–341.

McCarthy, James, and Scott Prudham. 2004. Neoliberal Nature and the Nature of Neoliberalism. *Geoforum* 35: 275–283.

McCay, Deirdre. 2005. Reading Remittance Landscapes: Female Migration and Agricultural Transition in the Philippines. *Geografisk Tidsskrift* 105(1): 89–99.

McDonald, David A., and Greg Ruiters. 2005. *The Age of Commodity: Water Privatization in Southern Africa*. London: Earthscan.

McDonald, Kevin. 2006. *Global Movements: Action and Culture*. Oxford: Blackwell Publishing.

McGranahan, Gordon. 2007. Urban Transitions and the Spatial Displacement of Environmental Burdens. In *Scaling Urban Environmental Challenges: From Local to Global and Back*, ed. Peter Marcotullio and Gordon McGranahan, 18–44. London: Earthscan.

McGranahan, Gordon, Deborah Balk, and Bridget Anderson. 2007. The Rising Tide: Assessing the Risks of Climate Change and Human Settlements in Low Elevation Coastal Zones. *Environment and Urbanization* 19(1): 17–37.

McGranahan, Gordon, Pedro Jacobi, Jacob Songsore, Charles Surjadi, and Marianne Kjellen. 2001. *The Citizens at Risk: From Urban Sanitation to Sustainable Cities*. London: Earthscan.

McGrew, Anthony, and Nana Poku. 2007. *Globalization, Development and Human Security*. Cambridge: Polity Press.

McInnes, Peter. 2005. Entrenching Inequalities: The Impact of Corporatization on Water Injustices in Pretoria. In *The Age of Commodity: Water Privatization in Southern Africa*, ed. David A. McDonald and Greg Ruiters, 99–119. London: Earthscan.

McLauchlan, Kendra. 2006. The Nature and Longevity of Agricultural Impacts on Soil Carbon and Nutrients: A Review. *Ecosystems* 9(8): 1364–1382.

McLean, R. F., S. K. Sinha, M. Q. Mirza, and M. Lal. 1998. Tropical Asia. In *The Regional Impacts of Climate Change: An Assessment of Vulnerability*, ed. R. T. Watson, M. C. Zinyowera, R. H. Moss, and D. J. Dokken, 383–407. Intergovernmental Panel on Climate Change. Cambridge: Cambridge University Press.

McMichael, A. J., C. D. Butler, and Carle Folke. 2003. New Visions for Addressing Sustainability. *Science* 302: 1919–1920.

Mearns, Linda O., Cynthia Rosenzweig, and Richard Goldberg. 1997. Mean and Variance Change in Climate Scenarios: Methods, Agricultural Applications, and Measures of Uncertainty. *Climatic Change* 35: 367–396.

Meehl, Gerald A., and Claudia Tebaldi. 2004. More Intense, More Frequent, and Longer Lasting Heat Waves in the 21st Century. *Science* 305: 994–997.

Meehl, Gerald A., Warren M. Washington, William D. Collins, Julie M. Arblaster, Aixue Hu, Lawrence E. Buja, Warren G. Strand, and Haiyan Teng. 2005. How Much More Global Warming and Sea Level Rise? *Science* 307 (March 18, 2005): 1769–1772.

Meikle, Sheilah. 2002. The Urban Context and Poor People. In *Urban Livelihoods: A People-Centred Approach to Reducing Poverty*, ed. Carole Rakodi and Tony Lloyd-Jones, 37–51. London: Earthscan.

Mellor, Mary. 2003. Gender and the Environment. In *Ecofeminism and Globalization: Exploring Culture, Context, and Religion*, ed. Heather Eaton and Lois Ann Lorentzen, 11–22. Lanham, MD: Rowman & Littlefield.

Mendelsohn, Robert. 2006. The Role of Markets and Governments in Helping Society Adapt to a Changing Climate. *Climatic Change* 78: 203–215.

Merchant, Carolyn. 2003. *Reinventing Eden: The Fate of Nature in Western Culture*. New York: Routledge.

Metz, Bert, Ogunlade Davidson, Heleen de Coninck, Manuela Loos, and Leo Meyer. 2005. *Carbon Dioxide Capture and Storage*. IPCC Special Report, Summary for Policymakers and Technical Summary. UNEP and WMO: Intergovernmental Panel on Climate Change.

Mies, Maria, and Vandana Shiva. 1993. *Ecofeminism*. Halifax, Nova Scotia: Fernwood Publications.

Milanovic, Branko. 2003. The Two Faces of Globalization: Against Globalization as We Know It. *World Development* 31(4): 667–683.

———. 2005. Half a World: Regional Inequality in Five Great Federations. *Journal of the Asia Pacific Economy* 10(4): 408–445.

Miles, Steven, and Malcolm Miles. 2004. *Consuming Cities*. New York: Palgrave MacMillan.

Mileti, Dennis. 1999. *Disasters by Design: A Reassessment of Natural Hazards in the United States*. Washington, DC: Joseph Henry.

Mileti, Dennis, and Julie L. Galius. 2005. Sustainable Development and Hazards Mitigation in the United States: Disasters by Design Revisited. *Mitigation and Adaptation Strategies for Global Change* 10(3): 491–504.

Millennium Ecosystem Assessment. 2003. *Ecosystems and Human Well-being: A Framework for Assessment*. Washington, DC: Island Press.

———. 2005. *Ecosystems and Human Well-being: Synthesis*. Washington, DC: Island Press.

Miller, Calum. 2005. *The Human Development Impact of Economic Crises*. United Nations, Human Development Report, Occasional Paper.

Miller, Scott. 2004. EU Reaches Pact on Farm Subsidies. *The Wall Street Journal Europe*. April 23–25, 2004.

Mills, Evan. 2005. Insurance in a Climate of Change. *Science* 309: 1040–1044.

Mills, Evan, Richard J. Roth, Jr., and Eugene Lecomte. 2005. *Availability and Affordability of Insurance under Climate Change: A Growing Challenge for the U.S.* Boston: Ceres.

Mitchell, Katharyne. 2004. *Crossing the Neoliberal Line: Pacific Rim Migration and the Metropolis*. Philadelphia: Temple University Press.

Mitchell, J. Kenneth, ed. 1999a. *Crucibles of Hazards: Megacities and Disasters in Transition*. Toyko: United Nations University Press.

———. 1999b. Megacities and Natural Disasters: A Comparative Analysis. *GeoJournal* 49(2): 137–142.

———. 2004. *Urbanization and Global Environmental Change: Integrative Science or Communicative Discourses?* International Conference on the Urban Dimensions of Environmental Change: Science, Exposures, Policies, and Technologies, Shanghai, June 25–28.

———. 2006. *The Globalization of Disaster Recovery*. Paper presented at Third Annual MaGrann Conference on "The Future of Disasters in a Globalizing World." New Brunswick, NJ: Rutgers University. April 21–22.

Mitra, A. P., D. Kumar, M. Rupa, K. Kumar, Y. P. Abrol, N. Kalra, M. Velayutham, and S. W. A. Naqvi. 2002. Global Change and Biogeochemical Cycles: The South Asia Region. In *Global-Regional Linkages in the Earth System*, ed. P. Tyson, R. Fuchs, C. Fu, L. Lebel, A. P. Mitra, E. Odada, J. Perry, W. Steffen, and H. Virji. Berlin: Springer.

Mittelman, James H. 1994. The Globalization Challenge: Surviving at the Margins. *Third World Quarterly*, 15: 427–443.

———. 2000. *The Globalization Syndrome: Transformation and Resistance*. Princeton, NJ: Princeton University Press.

Mittelman, James H., and Ashwini Tambe. 2000. Global Poverty and Gender. In *The Globalization Syndrome: Transformation and Resistance*, ed. James H. Mittelman, 74–89, Princeton, NJ: Princeton University Press.

Moe, Kjell A., and Gennady N. Semanov. 1999. Environmental Assessments. In *The Natural and Societal Challenges of the Northern Sea Route. A Reference Work*, ed. Willy Østreng, 121–219. Dordrecht: Kluwer Academic Publishers.

Mohan, G., E. Brown, B. Milward, and A. B. Zack-Williams. 2000. *Structural Adjustment: Theory, Practice and Impacts*. New York: Routledge.

Mol, Arthur. 2006. Environmental Governance in the Information Age: The Emergence of Informational Governance. *Environment and Planning C: Government and Policy* 24: 497–514.

Moser, Caroline O. N. 1998. The Asset Vulnerability Framework: Reassessing Urban Poverty Reduction Strategies. *World Development* 26(1): 1–19.

Moser, Susanne C., and Lisa Dilling, eds. 2007. *Creating a Climate for Change: Communicating Climate Change and Facilitating Social Change*. Cambridge: Cambridge University Press.

MSC. 2006. *MSC—Marine Stewardship Council Annual Report 2005/2006*. London: Marine Stewardship Council. Accessed June 15, 2007 <http://www.msc.org/assets/docs/MSC_annual_report_05_06.pdf>.

Mukherjee, Andy. 2005. In India, the Sky Sets Limits. *International Herald Tribune*, May 13, 2005.

Müller, Anders Riel, and Raj Patel. 2004. *Shining India? Economic Liberalization and Rural Poverty in the 1990s*. Policy Brief No. 10. Oakland: Food First Institute for Food and Development Policy.

Myers, Norman, and Jennifer Kent. 2003. New Consumers: The Influence of Affluence on the Environment. *Proceedings of the National Academy of Sciences (PNAS)* 100(8): 4963–4968.

Nader, Ralph, William Greider, Margaret Atwood, Vandana Shiva, Mark Ritchie, Wendell Berry, Jerry Brown, Herman Daly, Lori Wallach, Thea Lee, Martin Khor, David Phillips, Jorge Castañeda, Carlos Heredia, Davis Morris, and Jerry Mander. 1993. *The Case Against "Free Trade": GATT, NAFTA, and the Globalization of Corporate Power*. San Francisco: Earth Island Press.

Nakicenovic, Nebojsa, and Robert Swart, eds. 2000. *Special Report on Emissions Scenarios*. A Special Report of Working Group III of the Intergovernmental Panel on Climate Change. Cambridge: Cambridge University Press.

Narayanan, Sudha, and Ashok Gulati. 2002. *Globalization and the Smallholders: A Review of Issues, Approaches, and Implications*. MSSD (Markets and Structural Studies Division) Discussion Paper No. 50. Washington, DC: International Food Policy Research Institute and Rural Development Department, World Bank.

Neff, Robert Jon. 2007. *Transportation, Urban Development, and Greenhouse Gases: Patterns of Consumption and Justice in Philadelphia, Pennsylvania*. Ph.D. Dissertation, Department of Geography, Pennsylvania State University. University Park, PA.

Nelson, Valerie, Kate Meadows, Terry Cannon, John Morton, and Adrienne Martin. 2002. Uncertain Predictions, Invisible Impacts, and the Need to Mainstream Gender in Climate Change Adaptation. *Gender and Development* 10: 51–59.

Nepstad, Daniel C., Claudia M. Stickler, and Oriana Almeida. 2006. Globalization of the Amazon Soy and Beef Industries: Opportunities for Conservation. *Conservation Biology* 20: 1595–1603.

Newell, Barry, Carole L. Crumley, Nordin Hassan, Eric F. Lambin, Claudia Pahl-Wostl, Arild Underdal, and Robert Wasson. 2005. A Conceptual Template for Integrative Human–Environment Research. *Global Environmental Change* 15: 299–307.

Newman, Kathe, and Philip Ashton. 2004. Neoliberal Urban Policy and New Paths of Neighborhood Change in the American Inner City. *Environment and Planning A* **36**: 1151–1172.

Nikitin, Boris A., and Dilizhan A. Mirzoev. 1998. Complex Scientific and Technical Trends of Russian Federation Arctic Offshore Hydrocarbon Resource Development. *Ocean & Coastal Management* 41(2–3): 129–151.

Nilson, Artur, Andres Kiviste, Henn Korjus, Saadi Mihkelson, Ivar Etvark, and Tonu Oja. 1999. Impact of Recent and Future Climate Change on Estonian Foresty and Adaptation Tools. *Climate Research* 12: 205–214.

Norris, Fran H. 2005. *Range, Magnitude, and Duration of the Effects of Disasters on Mental Health: Review Update 2005.* Accessed February 27, 2007 <http://www.redmh.org/research/general/REDMH_effects.pdf>.

Northern Maritime Corridor. 2007. Welcome to the Northern Maritime Corridor, 2005–2008. Accessed June 15, 2007 <http://www.northernmaritimecorridor.no>.

Norwegian Ministry of Foreign Affairs. 2006. Regjeringens Nordområdestrategi [The Government's Northern Region Strategy]. Oslo, Norway. Accessed January 22, 2008 <http://www.regjeringen.no/upload/kilde/ud/pla/2006/0006/ddd/pdfv/302927-nstrategi06.pdf>.

NRC (National Research Council). 1992. *Global Environmental Change: Understanding the Human Dimensions.* Washington, DC: National Academy Press.

———. 1999. *Our Common Journey: A Transition Toward Sustainability.* Washington, DC: National Academy Press.

O'Brien, Karen L. 1998. *Sacrificing the Forest: Environmental and Social Struggles in Chiapas.* Boulder, CO: Westview Press.

O'Brien, Karen L., Jon Barnett, Indra de Soysa, Richard Matthew, Lyla Mehta, Joni Seager, Hans-Georg Bohle, and Maureen Woodrow. 2005. *Hurricane Katrina Reveals Challenges to Human Security.* AVISO 14, October 2005. Accessed June 15, 2007 <www.gechs.org/aviso/14.pdf>

O'Brien, Karen L., Siri Eriksen, Lynn Nygaard, and Ane Scholden. 2007. Beyond Semantics: Why Conceptualizations of Vulnerability Matter in Climate Change Discourses. *Climate Policy* 7: 73–88.

O'Brien, Karen L., and Robin M. Leichenko. 2000. Double Exposure: Assessing the Impacts of Climate Change within the Context of Economic Globalization. *Global Environmental Change* 10(3): 221–232.

———. 2003. Winners and Losers in the Context of Global Change. *Annals of the Association of American Geographers* 93(1): 89–103.

———. 2006. Climate Change, Equity and Human Security. *Die Erde* 137: 165–179.

———. 2007. *Human Security, Vulnerability and Sustainable Adaptation.* Background Paper for the UNDP Human Development Report 2007. New York: UNDP.

O'Brien, Karen L., Robin Leichenko, Ulka Kelkar, Henry Venema, Guro Aandahl, Heather Tompkins, Akram Javed, Suruchi Bhadwal, Stephan Barg, Lynn Nygaard, and Jennifer West. 2004. Mapping Vulnerability to Multiple Stressors: Climate Change and Globalization in India. *Global Environmental Change* 14(4): 303–313.

O'Brien, Karen L,. and Colleen Vogel, eds. 2003. *Coping with Climate Variability: The Use of Seasonal Climate Forecasts in Southern Africa.* Aldershot: Ashgate Publishing.

Odum, E. P. 1953. *Fundamentals of Ecology.* Philadelphia.: W. B. Saunders.

OECD. 1997. *Globalisation and Environment: Preliminary Perspectives.* Paris: Organisation for Economic Co-operation and Development.

————. 2005. *Handbook on Economic Globalization Indicators*. Paris: Organisation for Economic Co-operation and Development.

Ohmae, Kenichi. 1995. *The End of the Nation State: The Rise of Regional Economies*. New York: Free Press.

Olds, Kris. 2001. *Globalization and Urban Change: Capital, Culture, and Pacific Rim Mega-Projects*. New York: Oxford.

Oliver-Smith, Anthony. 2006. Disasters and Forced Migration in the 21st Century. In Social Science Research Council, *Understanding Katrina: Perspectives from the Social Sciences*. Accessed June 15, 2007 <http://understandingkatrina.ssrc.org/Oliver-Smith/>.

O'Loughlin, John, Lynn Staeheli, and Edward Greenberg. 2004. Globalization and Its Outcomes: An Introduction. In *Globalization and Its Outcomes*, ed. John O'Loughlin, Lynn Staeheli, and Edward Greenberg, 3–22. New York: Guilford Press.

O'Meara Sheehan, M. 2002. City Limits: Putting the Brakes on Sprawl. Worldwatch Paper 156, July 2002.

O'Neill, Karen M. 2006. *Rivers by Design; State Power and the Origins of U.S. Flood Control*. Durham, NC: Duke University Press.

Onis, Ziya, and Ahmet Faruk Aysan. 2000. Neoliberal Globalisation, the Nation-State and Financial Crises in the Semi-periphery: A Comparative Analysis. *Third World Quarterly* 21(1): 119–139.

Onis, Ziya, and Fikret Senses. 2005. Rethinking the Emerging Post-Washington Consensus. *Development and Change* 36: 263–290.

O'Riordon, Timothy, and Andrew Jordan. 1999. Institutions, Climate Change and Cultural Theory: Towards a Common Analytical Framework. *Global Environmental Change* 9: 81–93.

Osborne, Hillary. 2007. HSBC Pledges $100M to Combat Climate Change. *Guardian Unlimited* (May 30, 2007). Accessed May 31, 2007 <http://environment.guardian.co.uk/climatechange/story/0,,2091411,00.html>.

Østreng, Willy. 1999. The Historical and Geopolitical Context of the Northern Sea Route: Lessons to Be Considered. In *The Natural and Societal Challenges of the Northern Sea Route. A Reference Work*, ed. Willy Østreng. 1–46. Dordrecht: Kluwer Academic Publishers.

Palma, J. H. N., A. R. Graves, R. G. H. Bunce, P. J. Burgess, R. de Filippi, K. J. Keesman, H. van Keulen, F. Liagre, M. Mayus, G. Moreno, Y. Reisner, F. Herzog. 2007. Modeling Environmental Benefits of Silvoarable Agroforestry in Europe. *Agriculture, Ecosystems & Environment* 119(3/4): 320–334.

PAME. 2000. *Snap Shot Analysis of Maritime Activities in the Arctic*. Norwegian Maritime Directorate. Report No. 2000–3220.

Pande, Rekha. 2000. Globalization and Women in the Agricultural Sector. *International Feminist Journal of Politics* 2(3): 409–412.

Parry, Martin. 1990. *Climate Change and World Agriculture*. London: Earthscan.

Parry, Martin, Nigel Arnell, Tony McMichael, Robert Nicholls, Pim Partens, Sari Kovats, Matthew Livermore, Cynthia Rosenzweig, Ana Iglesias, and Gunther Fischer. 2001. Millions at Risk: Defining Critical Climate Change Threats and Targets. *Global Environmental Change* 11(3): 1–3.

Parry, Martin, Cynthia Rosenzweig, Ana Iglesias, Matthew Livermore, and Günther Fischer. 2004. Effect of Climate Change on Global Food Production under SRES Emissions and Socio-economic Scenarios. *Global Environmental Change* 14(1): 53–67.

Peck, Jamie. 1996. *Work Place: The Social Regulation of Labor Markets*. New York: Guilford Press.

Peck, Jamie, and Adam Tickell. 2002. Neoliberalizing Space. *Antipode* 34: 380–404.

Peet, Richard, and Michael Watts. 1996. Liberation Ecology: Development, Sustainability, and Environment in an Age of Market Triumphalism. In *Liberation Ecologies: Environment, Development, Social Movements*, ed. Richard Peet and Michael Watts, 1–45. London: Routledge.

Pelling, Mark. 2001. Natural Disaster? In *Social Nature: Theory, Practice, and Politics*, ed. Noel Castree and Bruce Braun, 170–188. Malden, MA: Blackwell Publishers.

———. 2003a. *The Vulnerability of Cities: Natural Disasters and Social Resilience*. London: Earthscan.

Pelling, Mark, ed. 2003b. *Natural Disasters and Development in a Globalizing World*. London: Routledge.

Peluso, Nancy L., and Michael Watts, eds. 2001. *Violent Environments*. Ithaca, NY: Cornell University Press.

Perrons, Diane. 2004. *Globalization and Social Change: People and Places in a Divided World*. London: Routledge.

Petrini, Carlo, and Benjamin Watson, eds. 2001. *Slow Food: Collected Thoughts on Taste, Tradition, and the Honest Pleasures of Food*. White River Junction, VT: Chelsea Green Publishing Company.

Piana, Grazia, and Yianna Lambrou. 2005. *Gender, the Missing Component in the Response to Climate Change*. Rome: FAO.

Pielke, Roger A., Jr., and Daniel Sarewitz. 2005. Bringing Society Back into the Climate Debate. *Population and Environment* 26(3): 255–268.

Pilkey, Orrin H., and J. Andrew G. Cooper. 2004. Society and Sea Level Rise. *Science* 303: 1781–1782.

Pingali, Prabhu. 2007. Westernization of Asian Diets and the Transformation of Food Systems: Implications for Research and Policy. *Food Policy* 32(3): 281–298.

Pinstrup-Andersen, Per. 2002. Food and Agricultural Policy for a Globalizing World: Preparing for the Future. *American Journal of Agricultural Economics* 84(5): 1201–1214.

Pirard, P., S. Vandentorren, M. Pascal, K. Laaidi, A. Le Tertre, S. Cassadou, and M. Ledrans. 2005. Summary of the Mortality Impact Assessment of the 2003 Heat Wave in France. *Euro Surveillance* 10(7):153–156.

Pitman, A. J. 2005. On the Role of Geography in Earth System Science. *Geoforum* 36: 137–148.

Plummer, Ryan, and Derek Armitage. 2007. A Resilience-Based Framework for Evaluating Adaptive Co-Management: Linking Ecology, Economics and Society in a Complex World. *Ecological Economics* 61: 62–74.

Polanyi, K. 1944. *The Great Transformation: The Political and Economic Origins of Our Time*. Boston: Beacon Press.

Polsky, Colin. 2004. Putting Space and Time in Ricardian Climate Change Impact Studies: Agriculture in the U.S. Great Plains, 1969–1992. *Annals of the Association of American Geographers* 94(3): 549–564.

Polsky, Colin, and David W. Cash. 2005. Drought, Climate Change, and Vulnerability: The Role of Science and Technology in a Multi-Scale, Multi-Stressor World. In *Drought and Water Crises: Science, Technology, and Management Issues*, ed. Donald A. Wilhite, 215–245. Boca Raton, FL: Taylor and Francis Group.

Polsky, Colin, Rob Neff, and Brent Yarnal. 2007. Building Comparable Global Change Vulnerability Assessments: The Vulnerability Scoping Diagram. *Global Environmental Change*, 17: 472–485.

Porter, Gareth. 1999. Trade Competition and Pollution Standards: "Race to the Bottom" or "Stuck at the Bottom"? *The Journal of Environment & Development* 8: 133–151.

Poumadère, Marc, Claire Mays, Sophie Le Mer, and Russel Blong. 2005. The 2003 Heat Wave in France: Dangerous Climate Change Here and Now. *Risk Analysis* 25(6): 1483–1494.

PPIAF. 2002. New Designs for Water and Sanitation Transactions: Making Private Sector Participation Work for the Poor, Report for the Public Private Infrastructure Advisory Facility (PPIAF) and Water and Sanitation Program (WSP). Accessed June 15, 2007 <http://www.wsp.org/filez/pubs/312200792648_NewDesigns4watsantransactions.pdf>.

Presas, L., and Melchert Saguas. 2004. Transnational Urban Spaces and Urban Environmental Reforms: Analyzing Beijing's Environmental Restructuring in Light of Globalization. *Cities* 21(4): 321–328.

Pretty, J. N, A. D. Noble, D. Bossio, J. Dixon, R. E. Hine, F. W. T. Penning De Vries, and J. I. L. Morison. 2006. Resource-Conserving Agriculture Increases Yields in Developing Countries. *Environmental Science & Technology* 40(4): 1114–1119.

Princen, Thomas, Michael Maniates, and Ken Conca. 2002. Confronting Consumption. In *Confronting Consumption, ed.* Thomas Princen, Michael Maniates, and Ken Conca, 1–20. Cambridge, MA: MIT Press.

Proshutinskly, Andrey Yu., Tatiana Proshutinsky, and Thomas Weingartner. 1998. *Environmental Conditions Affecting Commercial Shipping.* Final Report, INSROP Phase 2 Projects, Natural Conditions and Ice Navigation (Project 2, Box C: The Simulation of NSR Commercial Shipping). Accessed June 15, 2007 <http://www.ims.uaf.edu/insrop-2/report.html>.

Ragner, Claes Lykke, ed. 2000. *The 21st Century—Turning Point for the Northern Sea Route.* Dordrecht: Kluwer Academic Publishers.

Rahmstorf, Stefan. 2007. A Semi-Empirical Approach to Projecting Future Sea-Level Rise. *Science* 315: 368–370.

Rajan, R. S., and R. Sen. 2002. A Decade of Trade Reforms in India. *World Economics* 3: 87–100.

Rajan, Raghuram G., and Luigi Zingales. 2003. *Saving Capitalism from the Capitalists.* London: Random House Business Books.

Rakodi, Carole, and Tony Lloyd-Jones, eds. 2002. *Urban Livelihoods: A People-Centered Approach to Reducing Poverty.* London: Earthscan.

Rashid, Sabina Faiz. 2000. The Urban Poor in Dhaka City: Their Struggles and Coping Strategies during the Floods of 1998. *Disasters* 24(3): 240–253.

Rauscher, Michael. 1997. *International Trade, Factor Movements and the Environment.* Oxford: Clarendon Press.

Ravetz, Jerry. 2005. The Post-Normal Science of Safety. In *Science and Citizens*, ed. M. Leach, Ian Scoones, and Brian Wynne, 43–53. London: Zed Books.

Rawls, John. 1972. *A Theory of Justice.* Oxford: Oxford University Press.

Redclift, Michael, and Ted Benton, eds. 1994. *Social Theory and the Global Environment.* London: Routledge.

Redford, Kent, and J. Peter Brosius. 2006. Editorial: Diversity and Homogenization in the Endgame. *Global Environmental Change* 16: 317–319.

Reeves, Scott. 2005. Katrina's Impact Is Spreading. Forbes.com. August 31, 2005. Accessed January 11, 2007 <http://www.forbes.com/home/economy/2005/08/31/katrina-hurricane-impact-cx_sr_0831impact.html>.

Reilly, J., N. Hohmann, and S. Kane (1994). Climate Change and Agricultural Trade: Who Benefits, Who Loses? *Global Environmental Change* 4(1): 24–36.

Ribot, J. C., A. R. Magalhães, and S. S. Panagides, eds. 1996. *Climate Variability, Climate Change and Social Vulnerability in the Semi-arid Tropics*. Cambridge: Cambridge University Press.

Ritchie, Mark. 1993. Agricultural Trade Liberalization: Implications for Sustainable Agriculture. In *The Case Against "Free Trade": GATT, NAFTA, and the Globalization of Corporate Power*, ed. Ralph Nader et al., 163–194. Washington, DC: Island Press.

Ritzer, George. 2004. *The McDonaldization of Society*, rev. New Century ed. Thousand Oaks, CA: Pine Forge Press.

Rivlin, Gary. 2005. Storm and Crisis: Finances; After Two Storms, Cities Confront Economic Peril. *New York Times*, October 22, 2005. Accessed June 15, 2007 <http://select.nytimes.com/search/restricted/article?res=F10916F8385B0C718EDDA90994 DD404482>.

Robbins, Paul. 2004. *Political Ecology: A Critical Introduction*. Oxford: Blackwell Publishing.

Roberts, J. Timmons, and Bradley Parks. 2006. *A Climate of Injustice: Global Inequality, North-South Politics, and Climate Policy*. Cambridge, MA: MIT Press.

Roberts, J. Timmons, and Nikki Demetria Thanos. 2003. *Trouble in Paradise: Globalization and Environmental Crises in Latin America*. New York: Routledge.

Robinson, Guy. 2004. *Geographies of Agriculture: Globalisation, Restructuring and Sustainability*. London: Pearson/Prentice Hall.

Robinson, John. 2004. Squaring the Circle? Some Thoughts on the Idea of Sustainable Development. *Ecological Economics* 48: 369–384.

Rodrigue, Christine M. 2006. *Katrina/Rita and Risk Communication within FEMA*. Paper Presented at the 2006 Meeting of the Association of American Geographers, March 7–11, 2006, Chicago: Accessed June 15, 2007 <http://www.csulb.edu/~rodrigue/aag06katrina.html>.

Rodrik, Dani. 1997. *Has Globalization Gone Too Far?* Washington, DC: Institute for International Economics.

Rohde, David, Donald G. McNeil, Jr., Reed Abelson, and Shaila Dewan. 2005. 154 Patients Died, Many in Intense Heat, as Rescues Lagged. *New York Times*, September 19, 2005: 1.

Rosenzweig, Cyntia, and William D. Solecki, 2001. Global Environmental Change and a Global City: Lessons for New York. *Environment* 43(3): 8–18.

Roy, Marlene, and Henry David Venema. 2002. Reducing Risk and Vulnerability to Climate Change in India: The Capabilities Approach. *Gender and Development* 10(2): 78–83.

Rudel, Thomas K. 2002. Paths of Destruction and Regeneration: Globalization and Forests in the Tropics. *Rural Sociology* 67(4): 622–636.

———. 2005. *Tropical Forests: Regional Paths of Destruction and Regeneration in the Late Twentieth Century*. New York: Columbia University Press.

Ruitenbeek, Jack, and Cynthia Cartier. 2001. *The Invisible Wand: Adaptive Co-management as an Emergent Strategy in Complex Bio-economic Systems*. CIFOR Occasional Paper No. 34. Jakarta: Center for International Forestry Research (CIFOR).

Rupert, Mark. 2000. *Ideologies of Globalization: Contending Visions of a New World Order*. London: Routledge.

Rypdal, Kristin, Terje Berntsen, Jan S. Fuglestvedt, Asbjørn Torvanger, Kristin Aunan, Frode Stordal, and Lynn P. Nygaard. 2005. Tropospheric Ozone and Aerosols in Climate Agreements: Scientific and Political Challenges. *Environmental Science and Policy* 8(1): 29–43.

Sachs, Jeffrey D. 2005. *The End of Poverty: Economic Possibilities for Our Time*. New York: Penguin Press.

Sachs, Jeffrey D., Ashutosh Varshney, and Nirupam Bajpai. 2000. *India in the Era of Economic Reforms*. New York: Oxford University Press.

Sanchez, Omar. 2003. Globalization as a Development Strategy in Latin America? *World Development* 31(12): 1977–1995.

Sanchez-Rodriguez, Roberto, Karen C. Seto, David Simon, William D. Solecki, Frauke Kraas, and Gregor Laumann. 2005. *Science Plan: Urbanization and Global Environmental Change*. Bonn, Germany: International Human Dimensions Programme on Global Environmental Change.

Sassen, Saskia. 1991. *The Global City: New York, London, Toyko*. Princeton, NJ: Princeton University Press.

———. 1998. *Globalization and Its Discontents*. New York: The New Press.

Schär, C., P. L. Vidale, D. Lüthi, C. Frei, C. Häberli, M. A. Liniger, and C. Appenzeller. 2004. The Role of Increasing Temperature Variability in European Summer Heatwaves. *Nature* 427: 332–336.

Schellnhuber, Hans Joachim, Paul J. Crutzen, William C. Clark, Martin Claussen, and Hermann Held, eds. 2004. *Earth System Analysis for Sustainability*. Dahlem Workshop Reports. Cambridge, MA: MIT Press.

Schipper, Lisa, and Mark Pelling. 2006. Disaster Risk, Climate Change and International Development: Scope for, and Challenges to, Integration. *Disasters* 30: 19–38.

Schneider, Laura C. 2004. Bracken Fern Invasion in Southern Yucatan: A Case for Land Change Science. *The Geographical Review* 94(2): 229–241.

Schneider, Stephen H. 2001. A Constructive Deconstruction of Deconstructionists: A Response to Demeritt. *Annals of the Association of American Geographers* 91(2): 338–344.

Schofer, Evan, and Francisco J. Granados. 2006. Environmentalism, Globalization and National Economies, 1980–2000. *Social Forces* 85: 965–991.

Schröter, Dagmar, Colin Polsky, and Anthony G. Patt. 2004. Assessing Vulnerability to the Effects of Global Change: An Eight Step Approach. *Mitigation and Adaptation Strategies for Global Change* 10(4): 573–595.

Segnestam, Linda, Louise Simonsson, Jorge Rubiano, and Maria Morales. 2006. Cross-level Institutional Processes and Vulnerability to Natural Hazards in Honduras. Stockholm: Stockholm Environment Institute (SEI).

Sen, Amartya. 1999. *Development as Freedom*. New York: Anchor Books.

Shiva, Vandana. 2000. *Stolen Harvest: The Hijacking of the Global Food Supply*. New Delhi: India Research Press.

———. 2002. Globalization of Agriculture, Food Security, and Sustainability. In *Sustainable Agriculture and Food Security: The Impact of Globalization*, ed. Vandana Shiva and Gitanjali Bedi, 11–70. New Delhi: Sage Publications.

Shurmer-Smith, Pamela. 2000. *India. Globalization and Change*. London: Arnold.

Silva, Julie. 2007. Trade and Income Inequality in a Less Developed Country: The Case of Mozambique. *Economic Geography* 82(3): 111–136.

Silva, Julie, and Robin Leichenko. 2004. Regional Income Inequality and International Trade. *Economic Geography* 80: 261–286.

Sivakumar, M. V. K. 1998. Climate Variability and Food Vulnerability. *Global Change Newsletter (IGBP)* 35: 14–17.

Sklair, Leslie. 2002. *Globalization: Capitalism & Its Alternatives,* 3rd ed. Oxford: Oxford University Press.

Skoufias, Emmanuel. 2003. Economic Crises and Natural Disasters: Coping Strategies and Policy Implications. *World Development* 31(7): 1087–1102.

Skutch, Margaret M. 2002. Protocols, Treaties, and Action: The "Climate Change Process" Viewed through Gender Spectacles. *Gender and Development* 10(2): 30–39.

Slocum, Rachel. 2004. Consumer Citizens and the Cities for Climate Protection Campaign. *Environment and Planning A* 36: 763–782.

Smith, David. 1994. *Geography and Social Justice*. Cambridge, MA: Blackwell.

Smith, Joseph, and David Shearman. 2006. *Climate Change Litigation: Analysing the Law, Scientific Evidence, Health, and Property*. Adelaide, South Australia: Presidian.

Smith, L. C., G. M. MacDonald, A. A. Velichko, D. W. Beilman, O. K. Borisova, K. E. Frey, K. V. Kremenetski, and Y. Sheng. 2004. Siberian Peatlands a Net Carbon Sink and Global Methane Source Since the Early Holocene. *Science* 303: 353–356.

Smith, Mark J. 1998. *Social Science in Question: Towards a Postdisciplinary Framework*. London: Sage.

Smith, Michael Peter. 2001. *Transnational Urbanism: Locating Globalization*. Malden, MA: Blackwell.

Smith, Neil. 1984. *Uneven Development: Nature Capital and the Production of Space*. Oxford: Blackwell.

Soares-Filho, Britaldo Silveira, Daniel Curtis Nepstad, Lisa M. Curran, Gustavo Coutinho Cerqueira, Ricardo Alexandrino Garcia, Claudia Azevedo Ramos, Eliane Voll, Alice McDonald, Paur Lefebvre, and Peter Schlesinger. 2006. Modelling Conservation in the Amazon Basin. *Nature* 440: 520–523.

Solbrig, Otto T., Robert Paarlberg, and Francesco di Castri, eds. 2001. *Globalization and the Rural Environment*. Cambridge, MA: Harvard University Press.

Solecki, William D., and Robin M. Leichenko. 2006. Urbanization and the Metropolitan Environment: Lessons from New York and Shanghai. *Environment* 48(4): 8–23.

Solecki, William.D., and Cynthia Rosenzweig. 2007. Climate Change and Cities. Working Paper. The Institute for Sustainable Cities, City University of New York, Hunter College.

Sparks, Richard E. 2006. Rethinking, Then Rebuilding New Orleans. *Issues in Science and Technology* (Winter 2006). Accessed June 15, 2007 <http://www.issues.org/22.2/sparks.html>.

Speth, James Gustave, ed. 2003. *Worlds Apart: Globalization and the Environment*. Washington, DC: Island Press.

Steffen, W., A., P. D. Sanderson, Tyson, J. Jäger, P. A. Matson, M. Moore III, F. Oldfield, K. Richardson, H. J. Schellnhuber, B. L. Turner II, and R. J. Wasson. 2004. *Global Change and the Earth System: A Planet Under Pressure*. Berlin: Springer.

Stern, David I., Michael S. Common, and Edward B. Barbier. 1996. Economic Growth and Environmental Degradation: The Environmental Kuznets Curve and Sustainable Development. *World Development* 24: 1151–1160.

Stern, Nicholas. 2006. *The Economics of Climate Change: The Stern Review*. Cambridge: Cambridge University Press.

Stevens, Christopher. 2003. Food Trade and Food Policy in Sub-Saharan Africa: Old Myths and New Challenges. *Development Policy Review* 21(5–6): 669–681.

Stiglitz, Joseph E. 2002. *Globalization and Its Discontents*. New York: W.W. Norton.

———. 2006. *Making Globalization Work*. New York: W.W. Norton.

Stiglitz, Joseph E., and Andrew Charlton. 2005. *Free Trade for All: How Trade Can Promote Development*. Oxford: Oxford University Press.

Stokke, Olav Schram, and Geir Hønneland, eds. 2007. *International Cooperation and Arctic Governance: Regime Effectiveness and Northern Region Building*. London: Routledge.

Stokstad, Erik, et al. 2004. Defrosting the Carbon Freezing of the North. *Science* 304(1618): doi: 10.1126/science.304.5677.1618.

Storm, Servaas. 2003. Transition Problems in Policy Reform: Agricultural Trade Liberalization in India. *Review of Development Economics* 7(3): 406–418.

Stott, Peter A., D. A. Stone, and M. R. Allen. 2004. Human Contribution to the European Heatwave of 2003. *Nature* 432: 610–614.

Swart, Rob, John Robinson, and Stewart Cohen. 2003. Climate Change and Sustainable Development: Expanding the Options. *Climate Policy* 3S1: S19–S40.

Switzer, Jacqueline V. 1997. *Green Backlash: The History and Politics of the Environmental Movement in the U.S.* Boulder, CO: Lynne Rienner Publishers.

———. 2004. *Environmental Politics: Domestic and Global Dimensions,* 4th ed. Belmont, CA: Wadsworth/Thomson.

Tabachnick, David, and Toivo Koivukoski, eds. 2004. *Globalization, Technology, and Philosophy.* Albany: State University of New York Press.

Tan, Minghong, Xiubin Li, Hui Xie, and Changhe Lu. 2005. Urban Land Expansion and Arable Land Loss in China—A Case Study of Beijing-Tianjin-Hebei Region. *Land Use Policy* 22(3): 187–196.

Taylor, James M. 2005. Survey Shows Climatologists Are Split on Global Warming: Alarmist "Consensus" Does Not Exist. *Environment & Climate News* 8 (June): 14.

Taylor, Peter J., and Frederick H. Buttel. 1992. How Do We Know We Have Global Environmental Problems? Science and the Globalization of Environmental Discourse. *Geoforum* 23(3): 405–416.

Thompson, Alexander. 2006. Management under Anarchy: The International Politics of Climate Change. *Climatic Change* 78: 7–29.

Thompson, Alexander, Paul Robbins, Brent Sohngen, Joseph Arvai, and Tomas Koontz. 2006. Economy, Politics and Institutions: from Adaptation to Adaptive Management in Climate Change. *Climatic Change* 78: 1–5.

Tierney, Kathleen. 2006. Social Inequality, Hazards, and Disasters. In *On Risk and Disaster: Lessons from Katrina,* ed. Ronald J. Daniels, Donald F. Kettl, and Howard Kunreuther, 109–128. Philadelphia.: University of Pennsylvania Press.

Tilman, David, Joseph Fargione, Brian Wolff, Carla D'Antonio, Andrew Dobson, Robert Howarth, David Schindler, William H. Schlesinger, Daniel Simberloff, Deborah Swackhamer. 2001. Forecasting Agriculturally Driven Global Environmental Change. *Science* 292: 281–284.

Toffler, Alvin. 1970. *Future Shock.* New York: Random House.

Tol, Richard S. J., and Roda Verheyen. 2004. State Responsibility and Compensation for Climate Change Damages—A Legal and Economic Assessment. *Energy Policy* 32: 1109–1130.

Tol, Richard S. J., Maria Bohn, Thomas E. Downing, Marie-Laure Guillerminet, Eva Hizsnyik, Roger Kasperson, Kate Lonsdale, Claire Mays, Robert J. Nicholls, Alexander A. Olsthoorn, Gabriele Pfeifle, Marc Poumadere, Ferenc L. Toth, Athanasios T. Vafeidis, Peter E. van der Werff, and I. Hakan Yetkiner. 2006. Adaptation to Five Metres of Sea Level Rise. *Journal of Risk Research* 9(5): 467–482.

Trenberth, Kevin. 2005. Uncertainty in Hurricanes and Global Warming. *Science* 308: 1753–1754.

Turner, Billie L., II, William C. Clark, Robert W. Kates, John F. Richards, Jessica T. Mathews, and William B. Meyer. 1990. *The Earth as Transformed by Human Action. Global and Regional Changes in the Biosphere over the Past 300 Years.* Cambridge: Cambridge University Press with Clark University.

Turner, Billie L., II, Roger E. Kasperson, W. B. Meyer, Kirsten Dow, D. Golding, Jeanne X. Kasperson, R. C. Mitchell, and S. J. Ratick. 1991. Two Types of Global Environmental

Change: Definitional and Spatial Scale Issues in their Human Dimensions. *Global Environmental Change* 1(1): 14–22.

Turner, Billie L., II, Roger E. Kasperson, Pamela A. Matson, James J. McCarthy, Robert W. Corell, Lindsey Christensen, Noelle Eckley, Jeanne X. Kasperson, Amy Luers, Marybeth L. Martello, Colin Polsky, Alexander Pulsipher, and Andrew Schiller. 2003a. A Framework for Vulnerability Analysis in Sustainability Science. *Proceedings of the National Academy of Science (PNAS)* 100(14): 8074–8079.

Turner, Billie L., II, Pamela A. Matson, James J. McCarthy, Robert W. Corell, Lindsey Christensen, Noelle Eckley, Grete K. Hovelsrud-Broda, Jeanne X. Kasperson, Roger E. Kasperson, Amy Luers, Marybeth L. Martello, Svein Mathiesen, Rosamond Naylor, Colin Polsky, Alexander Pulsipher, Andrew Schiller, Henrik Selin, and Nicholas Tyler. 2003b. Illustrating the Coupled Human-Environment System for Vulnerability Analysis: Three Case Studies. *Proceedings of the National Academy of Science (PNAS)* 100(14): 8080–8085.

Turner, R. Eugene. 2004. Coastal Wetland Subsidence Arising from Local Hydrologic Manipulations. *Estuaries* 27(2): 265–273.

UN (United Nations). 2005a. *The Inequality Predicament: Report on the World Social Situation 2005.* New York: United Nations Department of Economic and Social Affairs.

———. 2005b. U.N. Millenium Development Goals. Accessed April 25, 2005 <http://www.un.org/millenniumgoals/index.html>.

———. 2006. *Atlas of Oceans.* Accessed March 31, 2007 <http://www.oceansatlas.org>.

UNDP (United Nations Development Programme). 2005. *Human Development Report 2005— International Cooperation at a Crossroads: Aid, Trade and Security in an Unequal World.* New York: United Nations Development Programme.

———. 2006. *Human Development Report 2006—Beyond Scarcity: Power, Poverty and the Global Water Crisis.* New York: United Nations Development Programme.

———. 2007. *Human Development Report 2007/2008—Fighting Climate Change: Human Solidarity in a Divided World.* New York: United Nations Development Programme.

UNEP (United Nations Environment Programme). 2006. *GEO Year Book 2006: An Overview of Our Changing Environment.* Nairobi: UNEP Division of Early Warning and Assessment.

UNIDO. 1995. *India: Towards Globalization.* Vienna: United Nations Industrial Development Organization.

U.S. Army Corps of Engineers, New Orleans Division. 1999. *Habitat Impacts of the Construction of the MRGO.* Report for the Environmental Subcommittee of the Technical Committee Convened by the Environmental Protection Agency (New Orleans, December 1999).

U.S. Census Bureau. 2005. American Community Survey. Washington, DC: Department of Commerce.

U.S. GAO. 2000. Aviation and the Environment: Aviation's Effects on the Global Atmosphere Are Significant and Expected to Grow. United States General Accounting Office, GAO/RCED 00-57. Accessed December 22, 2006 <http://www.gao.gov/archive/2000/rc00057.pdf>.

Vakulabharanam, Vamsi. 2005. Growth and Distress in a South Indian Peasant Economy During the Era of Economic Liberalisation. 2005. *The Journal of Development Studies* 41(6): 971–997.

Vale, Lawrence J., and Thomas J. Campanella, eds. 2005. *The Resilient City: How Modern Cities Recover from Disaster.* New York: Oxford University Press.

Van Heerden, Ivor, and Mike Bryan. 2006. *The Storm: What Went Wrong and Why During Hurricane Katrina—The Inside Story from One Lousiana Scientist.* New York: Viking Press.

van Meijl, H., T. van Rheenen, A. Tabeau, and B. Eickhout. 2006. The Impact of Different Policy Environments on Agricultural Land Use in Europe. *Agriculture, Ecosystems and Environment* 114: 21–38.

Van Vliet, Willem. 2002. Cities in a Globalizing World: From Engines of Growth the Agents of Change. *Environment and Urbanization* 14(1) 31–40.

Vásquez-Léon, Marcela, Colin Thor West, and Timothy J. Finan. 2003. A Comparative Assessment of Climate Vulnerability: Agriculture and Ranching on both Sides of the US-Mexico Border. *Global Environmental Change* 13: 159–173.

Vincent, Katherine. 2007. *Gendered Vulnerability to Climate Change in Limpopo Province, South Africa.* Ph.D. Dissertation. School of Development Studies, University of East Anglia, England.

Vogel, Coleen. 2005. "Seven Fat Years and Seven Lean Years"? Climate Change and Agriculture in Africa. *IDS Bulletin* 36(2): 30–35.

Vörösmarty, Charles. J. 2002. Global Water Assessment and Potential Contributions from Earth Systems Science. *Aquatic Sciences* 64: 328–351.

Vörösmarty, C. J., P. Green, J. Salisbury, and R. Lammers, 2000. Global Water Resources: Vulnerability from Climate Change and Population Growth. *Science* 289: 284–288.

Vörösmarty, C., D. Lettenmaier, C. Leveque, M. Meybeck, C. Pahl-Wostl, J. Alcamo, W. Cosgrove, H. Grassl, H. Hoff, P. Kabat, F.Lansigan, R.Lawford, and R. Naiman. 2004. Humans Transforming the Global Water System. *Eos Transactions, American Geophysical Union* 85(48): 509–520.

Wackernagel, M., and W. Rees. (1996). *Our Ecological Footprint: Reducing Human Impact on the Earth.* New Society Publishers, Philadelphia and Gabriola, Island, BC.

Walker, Brian, and David Salt. 2006. *Resilience Thinking: Sustaining Ecosystems and People in a Changing World.* Washington, DC: Island Press.

Wallerstein, Immanuel. 2000. Globalization or the Age of Transition: A Long-Term View of the Trajectory of the World-System. *International Sociology* 15: 249–265.

Walton, John, and David Seddon. 1994. *Free Markets and Food Riots: The Politics of Global Adjustment.* Cambridge, MA: Blackwell.

Wang, Ya Ping. 2001. Urban Housing Reform and Finance in China. *Urban Affairs Review* 36(5): 620–645.

WaterAid and Tearfund. 2003. *New Rules, New Roles: Does PSP Benefit the Poor?* Summaries of the Case Studies. London: WaterAid and Tearfund. Accessed June 15, 2007 <http://www.wateraid.org.uk/documents/plugin_documents/does_psp_benefit_the_poor.pdf>.

Watson, R. T., M. C. Zinyowera, R. H. Moss, and D. J. Dokken. 1996. *Climate Change 1995: Impacts, Adaptations and Mitigation of Climate Change: Scientific-Technical Analyses.* Cambridge: Cambridge University Press.

WCED (World Commission on Environment and Development). 1987. *Our Common Future.* Oxford: Oxford University Press.

Whelan, Robert K. 2006. An Old Economy for the "New" New Orleans? Post-Hurricane Katrina Economic Development Efforts. In *There Is No Such Thing as a Natural Disaster: Race, Class and Hurricane Katrina*, ed. Chester Hartman and Gregory D. Squires, 103–120. New York: Routledge.

Wilbanks, Thomas J., and Robert W. Kates. 1999. Global Change in Local Places: How Scale Matters. *Climatic Change* 43: 601–628.

Wilber, Ken. 2000. *A Brief History of Everything.* Dublin: Gateway.

Wilson, E. O., ed. 1988. *Biodiversity.* Washington: National Academies Press.

Wilson, Geoff A. 2001. From Productivism to Post-productivism...and Back Again? Exploring the (Un)changed Natural and Mental Landscapes of European Agriculture. *Transactions of the Institute of British Geographers* 26(1): 77–102.

Winters, L. Alan. 2004. Trade Liberalization and Economic Performance: An Overview. *The Economic Journal* 114 (Feb.): F4–F21.

Winton, Michael. 2006. Does the Arctic Sea Ice Have a Tipping Point? *Geophysical Research Letters* 33: L23504, doi: 10.1029/2006GL028017.

Wisner, Ben, Piers Blaikie, Terry Cannon, and Ian Davis. 2004. *At Risk: Natural Hazards, People's Vulnerability and Disasters*, 2nd ed. London: Routledge.

Wood, Geof. 2003. Staying Secure, Staying Poor: The "Faustian Bargain." *World Development* 31(3): 455–471.

Woodin, Michael, and Caroline Lucas. 2004. *Green Alternatives to Globalisation: A Manifesto.* London: Pluto Press.

World Bank. 2006a. *Privatizing Water and Sanitation Services.* Washington, DC: World Bank Accessed June 15, 2007 <http://rru.worldbank.org/PapersLinks/Privatizing-Water-Sanitation-Services/>.

———. 2006b. *World Development Report 2006: Equity and Development.* Washington, DC: World Bank.

World Social Forum India. 2007. Welcome to the World Social Forum India. New Delhi: World Social Forum India Office. Accessed May 30, 2007 <http://www.wsfindia.org/>.

Worm, Boris, Edward B. Barbier, Nicola Beaumont, J. Emmett Duffy, Carle Folke, Benjamin S. Halpern, Jeremy B. C. Jackson, Heike K. Lotze, Fiorenze Micheli, Stephen R. Palumbi, Enric Sala, Kimberley A. Selkoe, John J. Stachowicz, and Reg Watson. 2006. Impacts of Biodiversity Loss on Ocean Ecosystem Services. *Science* 314: 787–790.

Worster, Donald. 1985. *Rivers of Empire: Water, Aridity, and the Growth of the American West.* New York: Pantheon Books.

Wright, L., and L. Fulton. 2005. Climate Change Mitigation and Transport in Developing Nations. *Transport Reviews* 25(6): 691–717.

Wu, Fulong. 2004. Transplanting Cityscapes: The Use of Imagined Globalization in Housing Commodification in Beijing. *Area* 36: 227–243.

Wu, Fulong, and Klaire Webber. 2004. The Rise of "Foreign Gated Communities" in Beijing: Between Economic Globalization and Local Institutions. *Cities* 21(3): 203–213.

Wynne, Brian. 1994. Scientific Knowledge and the Global Environment. In *Social Theory and the Global Environment*, ed. Michael R. Redclift and Ted Benton, 169–189. London: Routledge.

Wysham, Daphne. 2005. A Carbon Rush at the World Bank. *Foreign Policy in Focus*, February 2005. Silver City, NM, and Washington, DC.

Yardley, William. 2007. Engulfed by Climate Change, Town Seeks Lifeline. *New York Times*, May 27, 2007. A1 and A25.

Yeoh, Brenda S. A. 1999. Global/Globalizing Cities. *Progress in Human Geography* 23(4): 607–616.

Young, Abby. 2007. Forming Networks, Enabling Leaders, Financing Action: The Cities for Climate Protection Campaign. In *Creating a Climate for Change: Communicating Climate Change and Facilitating Social Change*, ed. Susanne C. Moser and Lisa Dilling, 383–398. Cambridge: Cambridge University Press.

Young, Iris. 2006. Katrina: Too Much Blame, Not Enough Responsibility. *Dissent Magazine*, Winter 2006. Accessed June 15 <http://www.dissentmagazine.org/article/?article=158>.

Young, Oran. 2002. *The Institutional Dimensions of Environmental Change: Fit, Interplay, and Scale*. Cambridge, MA: MIT Press.

Young, Oran, Frans Berkout, Gilberto Gallopín, Marco A. Janssen, Elinor Ostrom, and Sander van der Leeuw. 2006. The Globalization of Socio-ecological Systems: An Agenda for Scientific Research. *Global Environmental Change* 16: 304–316.

Zarsky, Lyuba. 1997. Stuck in the Mud? Nation States, Globalisation, and Environment. In *Globalisation and Environment: Preliminary Perspectives*, 27–51. Paris: OECD Proceedings.

Zhang, Tingwei. 2000. Land Market Forces and the Government's Role in Sprawl: The Case of China. *Cities* 17(2): 123–135.

Zhang, Xing Quan. 2000. The Restructuring of the Housing Finance System in Urban China. *Cities* 17(5): 339–348.

Ziervogel, Gina, Sukaina Bharwani, and Thomas E. Downing. 2006. Adapting to Climate Variability: Pumpkins, People and Policy. *Natural Resources Forum* 30: 294–305.

Zimmerer, Karl. S. 2006. Geographical Perspectives on Globalization and Environmental Issues: The Inner-Connections of Conservation, Agriculture, and Livelihoods. In *Globalization & New Geographies of Conservation*, ed. Karl S. Zimmerer, 1–43. Chicago: University of Chicago Press.

Index